The ARRL 1987-1988 Technician/General Class License Manual for the Radio Amateur

Edited By

Larry D. Wolfgang, WA3VIL
Bruce S. Hale, KB1MW

Contributors

Mark J. Wilson, AA2Z
Jonathan F. Towle, WB1DNL
Leo D. Kluger, WB2TRN
Michael B. Kaczynski, W1OD

Production Staff

Deborah Strzeszkowski
David Pingree
Steffie Nelson, KA1IFB
Joel Kleinman, N1BKE
Sue Fagan, Cover Design
Michelle Chrisjohn, WB1ENT

American Radio Relay League
Newington, CT 06111 USA

Foreword

Welcome to the Third edition of the *ARRL Technician/General Class License Manual*. Major changes have occurred in Amateur Radio licensing and privileges since the last printing of this book. On January 28, 1987, acting on a request made by the ARRL and others, the FCC adopted a Report and Order that grants Novice class Amateur Radio operators many new operating privileges. Novice and Technician licensees have been granted digital and voice communications on 28 MHz.

Along with granting new privileges for Novices and Technicians, the FCC split the Element 3 exam, previously required for both Technician and General class licenses, into Elements 3A and 3B. The 3A exam is now required for the Technician license and the 3B exam is required for the General license. Applicants who wish to upgrade from Novice directly to General are responsible for *both* Element 3A and Element 3B. All of these rules changes took effect on March 21, 1987, and all exams given after that time follow these new procedures. The Technician and General class exams are each 25 questions under the new rules. Please note that if you have a Technician license issued before March 21, 1987, you automatically have credit for both Element 3A and Element 3B and only need to pass a 13 WPM code test to upgrade to General class.

This edition of the *ARRL Technician/General Class License Manual* was prepared to provide information about the new Element 3A and 3B exams. The complete Element 3A and Element 3B question pools appear in Chapters 10 and 11, along with the answer keys and page references to help you find the appropriate study material. We have expanded the references in the running text that direct you to study specific questions as they are covered in the text. The references now direct you not only to the appropriate questions, but to the particular pool you must study for the exam you will be taking.

Reference material for the questions on FCC regulations is found in the ARRL publication *The FCC Rule Book,* which contains the complete FCC Rules for Amateur Radio and full explanations to help you decipher the legalese. You should have a copy of that book as a companion to this one; together they provide all the information you need to pass the written tests.

League publications represent a real team effort. Writing, editing, typesetting, preparing technical illustrations and final layout all require special talents. The people performing those jobs put a lot of themselves into our publications to bring you the best possible books. Within our Publications Office the Technical Department editors and secretarial staff; the Production Department copy editors, technical illustrators, typesetters, layout staff and proofreaders; and the sales and shipping staffs of our Circulation Department have all been involved in helping you obtain the proper material to study for your Technician or General class exam. We couldn't possibly list all the names on the title page, but every one of us wishes you the very best on your exam. We think you will find the new privileges to be well worth the effort. We are proud of our work, and we hope you are proud to say you are a League member.

David Sumner, K1ZZ
Executive Vice President, ARRL
Newington, Connecticut

When To Use This Book

In mid-1987, a committee of Volunteer Examiner Coordinators (VECs) agreed on a three-year revision cycle for question pools from which Amateur Radio examinations are designed. Under the current schedule, the question pool for the Element 3A (Technician class) exam will be revised during 1988, and the pool for the Element 3B (General class) examination will be revised during 1989.

The revised Element 3A pool is scheduled to be released to VECs and publishers of Amateur Radio study material on February 1, 1989 for use in Technician class examinations administered beginning November 1, 1989. Therefore, this **Third Edition Technician-General License Manual may be used to prepare for Technician class exams administered before November 1, 1989.**

The revised Element 3B pool is scheduled to be released to VECs and publishers of Amateur Radio study material on February 1, 1990 for use in General class examinations administered beginning November 1, 1990. Therefore, this **Third Edition Technician-General License Manual may be used to prepare for General class exams administered before November 1, 1990.**

Do not use this book to study for Technician class exams administered after November 1, 1989 or General class exams administered after November 1, 1990.

Table of Contents

HOW TO USE THIS BOOK

To earn a Technician or General class Amateur Radio license, you will have to know some basic electronics theory and the rules and regulations governing the Amateur Radio Service, as contained in Part 97 of the FCC Rules. You'll also have to be able to send and receive the International Morse Code at a rate of 5 WPM for the Technician license and at 13 WPM for the General license. This book provides a brief description of the Amateur Radio Service, and the Technician and General class licenses in particular. In January 1987, the FCC split the Element 3 question pool (previously required for both General and Technician license candidates) into separate Element 3A and Element 3B pools. Applicants for the Technician license must pass a 25-question exam drawn from the questions in Element 3A, and applicants for the General license must pass a 25-question exam drawn from Element 3B. Please note that Novice licensees (or applicants without an amateur license) wishing to upgrade directly to General class must pass *both* the Element 3A and the Element 3B exams. Chapters 10 and 11 contain the complete Element 3A and Element 3B question pools and multiple-choice answers that will be used by many Volunteer-Examiner Coordinators (including the ARRL/VEC) throughout 1987.

At the beginning of each chapter, you will find a list of "key words" that appear in that chapter, along with a simple definition for each word or phrase. As you read the text, you will find these words printed in *italic* type the first time they appear. You may want to refer back to the beginning of the chapter at that point, to review the definition. After you have studied the material in that chapter, you may also want to go back and review those definitions. Since most of the key words are terms that will appear on exam questions, a quick review of the key words list for each chapter just before you go to take the test will also be helpful.

As you study the material, you will be instructed to turn to sections of the questions in Chapter 10 and Chapter 11. Be sure to use these questions to review your understanding of the material at the suggested times. This will break the material into "bite-sized pieces," and make it easier for you to learn. Do not try to memorize all of the questions. That will be impossible! Instead, by using the questions for review, you will be familiar with them when you take the test, but you will also understand the electronics theory behind the questions. Remember to study the questions in *both* Chapter 10 and Chapter 11 if you are upgrading directly from Novice (or no license) to General.

In addition to this book, you will want to purchase a copy of the ARRL publication, *The FCC Rule Book*, which covers all the rules and regulations you'll need to know. If you are planning to take the General class exam you will probably need to increase your code speed, and ARRL also offers a complete set of cassette tapes to help you build up your code speed. Even with the tapes, you'll want to tune in to the code-practice sessions transmitted by W1AW, the ARRL Headquarters station. When you are almost able to copy the code at 13 WPM, you may find it helpful to listen to code at 15 WPM at the beginning of your practice session, and then decrease the speed to 13 WPM. The W1AW fast-code-practice sessions start at 35 WPM and decrease to 30, 25 and 20 WPM before getting to 15- and 13-WPM practice, so you can tune in to the practice session a little late and still get the practice you need. You might also want to try copying the W1AW CW bulletins, which are sent at 18 WPM. For more information about W1AW or how to order any ARRL publication or code tape, write to: ARRL Headquarters, 225 Main Street, Newington, CT 06111.

Chapter 1

The Technician and General Class Licenses

M
any people regard the Technician license as a good starting point in Amateur Radio, and some prefer to earn the General license as their first "ticket." For those who start out at the Novice level, one of these licenses is considered to be a giant step in amateur privileges. After all, with a Technician license comes the privilege of operating through repeaters on the popular 2-meter band, and the General class license grants high-power phone privileges on the amateur HF bands.

If you have put off earning your Technician or General license because you thought the electronics theory was too difficult, or the 13-WPM code speed was an impossible goal, put those thoughts aside right now! Once you make the commitment to study and learn what it takes to pass the exam, you *will* be able to do it. It often takes more than one try to pass the exam, but many amateurs do pass on their first try. The key is that you must make the commitment, and be willing to study. To build your code speed there are many good Morse code training techniques, including the ARRL code tapes, W1AW code practice and even some computer programs. So with this book, carefully designed to teach the required electronics theory, *The FCC Rule Book* published by the ARRL, and plenty of code practice, you will soon have a higher-class license. Whether you now hold a Novice license, or even if you don't yet have any license, you will find the operating privileges available to a Technician or General class licensee to be worth the time spent learning about your hobby.

With a Novice or Technician class license, you can "work the world" on segments of four different high-frequency (HF) bands, using the ham's special language, international Morse code. In addition, Technicians may operate on all frequency bands allocated to the Amateur Radio Service above the HF bands (and there are a lot of them!), ranging from the 6-meter and popular 2-meter bands to the upper reaches of the frequency spectrum, where amateur work is largely experimental. The recent "Novice Enhancement" proceeding allows Novice and Technician licensees to use single-sideband voice and digital modes on the 10-meter band.

General class operators are allowed the use of substantially larger portions of the HF bands that Novices and Technicians use, as well as several other bands in the HF range. Worldwide communication through both Morse code and voice operation (as well as modes such as slow-scan TV, radioteletype and facsimile) is available to General class licensees.

You can see that the Technician and General class licenses allow a wide range of transmitting privileges. It is no wonder that the General class license is the most popular one in the US. To earn those privileges, however, you will have to demonstrate that you know the international Morse code, basic electronic theory, operating practices, and FCC rules and regulations. This book is designed to teach you what you need to know to qualify for the Technician or General class Amateur Radio license.

IF YOU'RE A NEWCOMER TO AMATEUR RADIO

Earning an Amateur Radio license, at whatever level, is a special achievement. The half a million or so people in the US who call themselves Amateur Radio operators, or hams, are part of a global fraternity. Radio amateurs serve the public as a voluntary, noncommercial, communication service. This is especially true during natural disasters or other emergencies. In addition, hams continue to make important contributions to the field of electronics. Amateur Radio experimentation is yet another reason many people become part of this self-disciplined group of trained operators, technicians and electronics experts—an asset to any country. Hams pursue their hobby purely for personal enrichment in technical and operating skills, without considera-tion of any type of payment.

Because radio signals do not know territorial boundaries, hams have a unique ability (and responsibility) to enhance international goodwill. A ham becomes an ambassador of his country every time he puts his station on the air.

Amateur Radio has been around since before World War I, and hams have always been at the forefront of technology. Today hams relay signals through their own satellites in the OSCAR (Orbiting Satellite Carrying Amateur Radio) series, bounce signals off the moon, and use any number of other exotic communications techniques. Amateurs talk from hand-held transceivers through mountaintop repeater stations that can relay their signals to transceivers in other hams' cars or homes. Hams send pictures by television, chat with other hams around the world by voice or tap out messages in Morse code. When emergencies arise, radio amateurs are on the spot to relay information to and from disaster-stricken areas that have lost normal lines of communication.

The US government, through the Federal Communications Commission (FCC), grants all US Amateur Radio licenses. This licensing procedure ensures that radio amateurs possess the necessary operating skill and electronics knowledge. Without these skills, radio operators might unknowingly cause interference to other services using the radio spectrum because of improperly adjusted equipment or neglected regulations.

Who Can Be a Ham?

The FCC doesn't care how old you are or whether you're a US citizen: If you pass the examination, the Commission will issue you an amateur license. Any person (except the agent of a foreign government) may take an exam, and, if successful, receive an amateur license. It's important to understand that if a citizen of a foreign country receives an amateur license in this manner, he or she is a US Amateur Radio operator. (This should not be confused with reciprocal licensing, which allows visitors from certain countries who hold valid amateur licenses in their homelands to operate their own stations in the US without having to take an FCC exam.)

Licensing Structure

By examining Table 1-1, you'll see that there are five amateur license classes. Each class has its own requirements and privileges. The FCC requires proof of your ability to operate an amateur station properly. The required knowledge is in line with the privileges of the license you hold. Higher license classes require more knowledge— and offer greater operating privileges. The specific operating privileges for Technician and General class licensees are listed in Table 1-2. So as you upgrade your license class, you must pass more challenging written examinations.

In addition, you must demonstrate an ability to receive international Morse code

Table 1-1

Amateur Operator Licenses†

Class	Code Test	Written Examination	Privileges
Novice	5 WPM (Element 1A)	Elementary theory and regulations (Element 2)	Telegraphy in 3700-3750, 7100-7150, and 21,100-21,200 kHz with 200 watts PEP output maximum; telegraphy and RTTY on 28,100-28,300 kHz and telegraphy and SSB voice on 28,300-28,500 kHz with 200 W PEP max; all amateur modes authorized on 222.1-223.91 MHz, 25 W PEP max; all amateur modes authorized on 1270-1295 MHz, 5 W PEP max.
Technician	5 WPM (Element 1A)	Elementary theory and regulations; general-level theory and regulations. (Elements 2 and 3A)	All amateur privileges above 50.0 MHz plus Novice privileges.
General	13 WPM (Element 1B)	Elementary theory and regulations; general theory and regulations. (Elements 2, 3A and 3B)	All amateur privileges except those reserved for Advanced and Extra Class; see Table 1-2
Advanced	13 WPM (Element 1B)	General theory and regulations, plus intermediate theory. (Elements 2, 3A, 3B and 4A)	All amateur privileges except those reserved for Amateur Extra Class; see Table 1-2.
Amateur Extra Class	20 WPM (Element 1C)	General theory and regulations, intermediate theory, plus special exam on advanced techniques. (Elements 2, 3A, 4A and 4B)	All amateur privileges

†A licensed radio amateur will be required to pass only those elements that are not included in the examination for the amateur license currently held.

at 5 WPM for Novice and Technician, 13 WPM for General and Advanced, and 20 WPM for Amateur Extra. It's important to stress that although you may intend to use voice rather than code, this doesn't excuse you from the code test. By international treaty, knowing the international Morse code is a basic requirement for operating on any amateur band below 30 MHz.

Learning the Morse code is a matter of practice. Instructions on learning the code, how to handle a telegraph key, and so on can be found in the ARRL *Tune in the World with Ham Radio* package. This package includes two code-teaching cassettes for beginners. Additional cassettes for code practice at speeds of 5, 7½, 10, 13, 15 and 20 WPM are available from the American Radio Relay League, Newington, CT 06111.

Code Practice

Besides listening to code tapes, some on-the-air operating experience will be a great help in building your code speed. When you are in the middle of a contact via Amateur Radio, and have to copy the code that the other station is sending to continue the conversation, your copying ability will improve quickly!

ARRL's Hiram Percy Maxim Memorial Station, W1AW, transmits code practice and bulletins of information of interest to all amateurs. These code-practice sessions and Morse code bulletins provide an excellent opportunity for code practice. Table 1-3 is an abbreviated W1AW operating schedule. When we change from Standard Time to Daylight Saving Time, the same local times are used.

Table 1-2
Amateur Operating Privileges

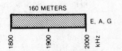

160 METERS — E, A, G

1800 1900 2000 kHz

Amateur stations operating at 1900-2000 kHz must not cause harmful interference to the radiolocation service and are afforded no protection from radiolocation operations; see January 1986 Happenings for details.

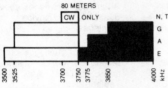

80 METERS

CW ONLY

N, T
G
A
E

3500 3525 3700 3750 3775 3850 4000 kHz

5167.5 kHz Alaska emergency use only. (SSB only)

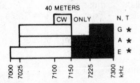

40 METERS

CW ONLY

N, T
G ★
A ★
E ★

7000 7025 7100 7150 7225 7300 kHz

* Phone operation is allowed on 7075-7100 kHz in Puerto Rico, US Virgin Islands and areas of the Caribbean south of 20 degrees north latitude; and in Hawaii and areas near ITU Region 3, including Alaska.

30 METERS — E, A, G

10,100 10,150 kHz

Maximum power limit on 30 meters is 200 watts PEP output. Amateurs must avoid interference to the fixed service outside the US.

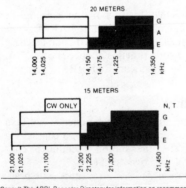

20 METERS

G
A
E

14,000 14,025 14,150 14,175 14,225 14,350 kHz

15 METERS

CW ONLY

N, T
G
A
E

21,000 21,025 21,100 21,200 21,225 21,300 21,450 kHz

Consult *The ARRL Repeater Directory* for information on recommended operating frequencies and band plans for the bands above 50 MHz.

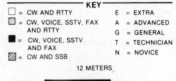

KEY

☐ = CW AND RTTY
▨ = CW, VOICE, SSTV, FAX AND RTTY
■ = CW, VOICE, SSTV AND FAX
▩ = CW AND SSB

E = EXTRA
A = ADVANCED
G = GENERAL
T = TECHNICIAN
N = NOVICE

12 METERS — E, A, G

24,890 24,930 24,990 kHz

Amateurs must avoid interference to the fixed service outside the US.

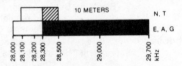

10 METERS

N, T
E, A, G

28,000 28,100 28,200 28,300 28,500 29,000 29,700 kHz

Novices and Technicians are limited to 200 watts on 10 meters.

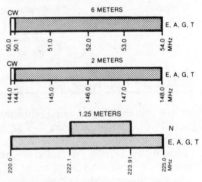

6 METERS

CW — E, A, G, T

50.0 50.1 51.0 52.0 53.0 54.0 MHz

2 METERS

CW — E, A, G, T

144.0 144.1 145.0 146.0 147.0 148.0 MHz

1.25 METERS

N
E, A, G, T

220.0 222.1 223.91 225.0 MHz

Novices are limited to 25 watts PEP on 222.1 to 223.91 MHz

70 CENTIMETERS — E, A, G, T

420.0 450.0 MHz

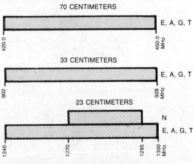

33 CENTIMETERS — E, A, G, T

902 928 MHz

23 CENTIMETERS

N
E, A, G, T

1240 1270 1295 1300 MHz

Novices are limited to 5 watts PEP on 1270 to 1295 MHz.

Table 1-3
W1AW Schedule

MTWThFSSn = Days of Week   Dy = Daily

W1AW code practice and bulletin transmissions are sent on the following schedule:

EST	Slow Code Practice	MWF: 9 AM, 7 PM; TThSSn: 4 PM, 10 PM
	Fast Code Practice	MWF: 4 PM, 10 PM; TTh: 9 AM; TThSSn: 7 PM
	CW Bulletins	Dy: 5 PM; 8 PM, 11 PM; MTWThF: 10 AM
	Teleprinter Bulletins	Dy: 6 PM, 9 PM, 12 PM; MTWThF: 11 AM
	Voice Bulletins	Dy: 9:30 PM, 12:30 AM
CST	Slow Code Practice	MWF: 8 AM, 6 PM; TThSSn: 3 PM, 9 PM
	Fast Code Practice	MWF: 3 PM, 9 PM; TTh: 8 AM; TThSSn: 6 PM
	CW Bulletins	Dy: 4 PM, 7 PM, 10 PM; MTWThF: 9 AM
	Teleprinter Bulletins	Dy: 5 PM, 8 PM, 11 PM; MTWThF: 10 AM
	Voice Bulletins	Dy: 8:30 PM, 11:30 PM
MST	Slow Code Practice	MWF: 7 AM, 5 PM; TThSSn: 2 PM, 8 PM
	Fast Code Practice	MWF: 2 PM, 8 PM; TTh: 7 AM; TThSSn: 5 PM
	CW Bulletins	Dy: 3 PM, 6 PM, 9 PM; MTWThF: 8 AM
	Teleprinter Bulletins	Dy: 4 PM, 7 PM, 10 PM; MTWThF: 9 AM
	Voice Bulletins	Dy: 7:30 PM, 10:30 PM
PST	Slow Code Practice	MWF: 6 AM, 4 PM; TThSSn: 1 PM; 7 PM
	Fast Code Practice	MWF: 1 PM, 7 PM; TTh: 6 AM; TThSSn: 4 PM
	CW Bulletins	Dy: 2 PM, 5 PM, 8 PM; MTWThF: 7 AM
	Teleprinter Bulletins	Dy: 3 PM, 6 PM, 9 PM; MTWThF: 8 AM
	Voice Bulletins	Dy: 6:30 PM, 9:30 PM

Code practice, Qualifying Run and CW bulletin frequencies: 1.818, 3.58, 7.08, 14.07, 21.08, 28.08, 50.08, 147.555 MHz.

Teleprinter bulletin frequencies: 3.625, 7.095, 14.095, 21.095, 28.095, 147.555 MHz.
Voice bulletin frequencies: 1.89, 3.99, 7.29, 14.29, 21.39, 28.59, 50.19, 147.555 MHz.

On Monday, Wednesday and Friday, 9 AM through 5 PM EST, transmissions are beamed to Europe on 14, 21 and 28 MHz; on Wednesday at 6 PM EST they are beamed south.

Slow code practice is at 5, 7½, 10, 13 and 15 WPM. Fast code practice is at 35, 30, 25, 20, 15, 13 and 10 WPM.

Code practice texts are from QST, and the source of each practice is given at the beginning of each practice and at the beginning of alternate speeds. For example, "Text is from July 1985 QST, pages 9 and 76" indicates that the main text is from the article on page 9 and the mixed number/letter groups at the end of each speed are from the contest scores on page 76.

Some of the slow practice sessions are sent with each line of text from QST reversed. For example, "Last October, the ARRL Board of Directors" would be sent as DIRECTORS OF BOARD ARRL THE, OCTOBER LAST.

W1AW CW and voice bulletins are sent on OSCAR 10, Mode B, when the satellite is within range. Look for CW on 145.840 MHz and SSB on 145.962 MHz.

Teleprinter bulletins are 45.45-baud Baudot, 110-baud ASCII and 100-baud AMTOR, FEC mode. Baudot, ASCII and AMTOR (in that order) are sent during all 11 AM EST transmissions, and 6 PM EST on TThFSSn. During other transmission times, AMTOR is sent only as time permits.

CW bulletins are sent at 18 WPM.

W1AW is open for visitors Monday through Friday from 8 AM to 1 AM EST and on Saturday and Sunday from 3:30 PM to 1 AM EST. If you desire to operate W1AW, be sure to bring a copy of your license with you. W1AW is available for operation by visitors between 1 and 4 PM Monday through Friday.

In a communications emergency, monitor W1AW for special bulletins as follows: voice on the hour, teleprinter at 15 minutes past the hour, and CW on the half hour.

Station Call Signs

Many years ago, by international agreement, the nations of the world decided to allocate certain call-sign prefixes to each country. This means that if you hear a radio station call sign beginning with W or K, for example, you know the station is licensed by the United States. A call sign beginning with the letter U is licensed by the USSR, and so on.

International Telecommunication Union (ITU) radio regulations outline the basic principles used in forming amateur call signs. According to these regulations, an amateur call sign must be made of one or two characters (either of which may be a numeral) as a prefix, followed by a numeral, and then a suffix of not more than three letters. The prefixes W, K, N and A are used in the United States. The continental US is divided into 10 Amateur Radio call districts (sometimes called areas), numbered 0 through 9. Figure 1-1 is a map showing the US call districts.

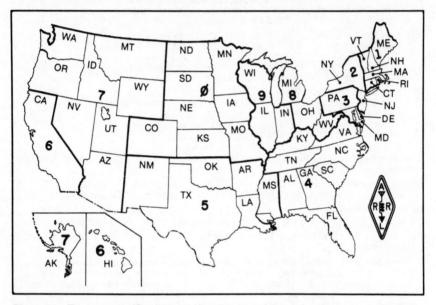

Figure 1-1—There are 10 US call areas. Hawaii is part of the sixth call area, and Alaska is part of the seventh.

All US Amateur Radio call signs assigned by the FCC after March 1978 can be categorized into one of five groups, each corresponding to a class, or classes, of license. Call signs are issued systematically by the FCC; requests for special call signs are not granted. For further information on the FCC's call-sign assignment system, and a table listing the blocks of call signs for each license class, see the ARRL publication, *The FCC Rule Book*. If you already have an amateur call sign, you may keep the same one when you change license class, if you wish. You must indicate your preference to receive a new call sign when you fill out an FCC Form 610 to apply for the exam or change your address.

EARNING A LICENSE
Applying for an Exam: FCC Form 610

Before you can take an FCC exam, you'll have to fill out a Form 610. This form is used as an application for a new license, an upgraded license, a renewal of a license or a modification to a license. In addition, hams who have held a valid license that has expired within the past two years may apply for reinstatement (and their old call sign) with a Form 610. Amateurs whose license expired more than two years ago but less than five years ago may also apply for reinstatement of their license, but they will be issued a new call sign. The FCC has a set of detailed instructions for the Form 610. These instructions are printed in *The FCC Rule Book* published by the ARRL. Form 610s showing a revision date before June 1984 are no longer valid, and will be returned if you attempt to use one. If you have any old forms, throw them away. To obtain a new Form 610, send a business-size, self-addressed, stamped envelope to: Form 610, ARRL, 225 Main Street, Newington, CT 06111.

Volunteer-Examiner Program

Since January 1, 1985, all US amateur exams above the Novice level have been administered under the Volunteer-Examiner Program. *The FCC Rule Book* contains details on this program. Novice exams have always been given by volunteer examiners, and that is still true. Those exams do not come under the regulations involving Volunteer-Examiner Coordinators (VECs), however.

To qualify for a Technician class license, you must pass FCC Exam Element 1A, Element 2 and Element 3A. For the General class license you will also be required to pass Exam Element 1B and Element 3B. If you already hold a valid Novice license, then you have credit for passing Elements 1A and 2, and will not have to retake them when you go for your Technician or General class exam. See Table 1-1 for details.

The FCC specifies the minimum number of questions for an Element 3A and Element 3B exam, along with a percentage of questions from each subelement that must appear on the exam. Most Element 3A and Element 3B exams (including exams coordinated by the ARRL/VEC) consist of 25 questions. The questions are taken from a pool of more than 250 questions, published in advance.

Volunteer-Examiner Coordinators are now responsible for maintaining the pools of questions that are used on amateur exams, and most VECs, including the ARRL/VEC, have agreed to use the question pools printed in Chapter 10 and Chapter 11 of this book. If your test session is coordinated by the ARRL/VEC or one of the other VECs who have agreed to use these question pools, the questions and answers on your exam will appear just as they do in this book. Some VECs may use the same questions with different answers; check with the VEC coordinating the test session you plan to attend.

As you study the questions in the question pools, remember that you are responsible for the questions in *both* Element 3A and Element 3B if you are upgrading directly from Novice (or no license) to General class. If you are upgrading from Novice (or no license) to Technician, study the questions in Element 3A. If you have a Technician license issued after March 21, 1987, and you wish to upgrade to General, you must study the questions in Element 3B. Finally, if you have a Technician license issued *before* March 21, 1987, you will be given credit for both Element 3A and Element 3B, and you only need to pass the 13 WPM code exam to upgrade to General class.

Finding an Exam Opportunity

To determine where and when an exam will be given, contact the ARRL/VEC Office, or watch for announcements in the Hamfest Calendar and Coming Conventions columns in *QST*. Many local clubs sponsor exams, so they are another good source of information on exam opportunities. ARRL officials such as Directors, Vice Directors and Section Managers receive notices about test sessions in their area. See page 8 in the latest issue of *QST* for names and addresses.

To register for an exam, send a completed Form 610 to the Volunteer-Examiner

Team responsible for the exam session if preregistration is required. Otherwise, bring the form to the session. Registration deadlines, and the time and location of the exams, are mentioned prominently in publicity releases about upcoming sessions.

Taking The Exam

By the time examination day rolls around, you should have already prepared yourself. This means getting your schedule, supplies and mental attitude ready. Plan your schedule so you'll get to the examination site with plenty of time to spare. There's no harm in being early. In fact, you might have time to discuss hamming with another applicant, which is a great way to calm pretest nerves. Try not to discuss the material that will be on the examination, as this may make you even more nervous. By this time, it's too late to study anyway!

What supplies will you need? First, be sure you bring your current *original* Amateur Radio license, if you have one. Bring along several sharpened number 2 pencils and two pens (blue or black ink). Be sure to have a good eraser. A pocket calculator will also come in handy. You may use a programmable calculator if that is the kind you have, but take it into your exam "empty" (cleared of all programs and constants in memory). Don't program equations ahead of time, because you may be asked to demonstrate that there is nothing in the calculator's memory. The Volunteer-Examiner Team is required to check two forms of identification before you enter the test room. This includes your *original* Amateur Radio license. A photo ID of some type is best for the second form, but is not required by FCC. Other acceptable forms of identification include a driver's license, a piece of mail addressed to you, a birth certificate or some other such document.

Before you take the code test, you'll be handed a piece of paper to copy the code as it's sent. The test will begin with about a minute of practice copy. Then comes the actual test: five minutes of Morse code. You are responsible for knowing the 26 letters of the alphabet, the numerals 0 through 9, the period, comma, question mark, $\overline{AR}$, $\overline{SK}$, $\overline{BT}$ and $\overline{DN}$. You may copy the entire text word for word, or just take notes on the content. At the end of the transmission, the examiner will hand you 10 questions about the text. Simply fill in the blanks with your answers. (You must spell each answer exactly as it was sent.) If you get at least 7 correct, you pass! Alternatively, the exam team has the option to look at your copy sheet if you fail the 10-question exam. If you have one minute of solid copy, they can certify that you passed the test on that basis. The format of the test transmission is similar to one side of a normal on-the-air amateur conversation.

A sending test may not be required. The Commission has decided that if applicants can demonstrate receiving ability, they most likely can also send at that speed. But be prepared for a sending test, just in case!

If all has gone well with the code test, you'll then take the written examination. The examiner will give each applicant a test booklet, an answer sheet and scratch paper. After that, you're on your own. The first thing to do is read the instructions. Be sure to sign your name every place it's called for. Do all of this at the beginning to get it out of the way.

Next, check the examination to see that all pages and questions are there. If not, report this to the examiner immediately. When filling in your answer sheet, make sure your answers are marked next to the numbers that correspond to each question.

Go through the entire exam, and answer the easy questions first. Next, go back to the beginning and try the harder questions. The really tough questions should be left for last. Guessing can only help, as there is no additional penalty for answering incorrectly.

If you have to guess, do it intelligently: At first glance, you may find that you can eliminate one or more "distractors." Of the remaining responses, more than one may seem correct; only one is the best answer, however. To the applicant who is fully prepared, incorrect distractors to each question are obvious. Nothing beats preparation!

After you've finished, check the examination thoroughly. You may have read a question wrong or goofed in your arithmetic. Don't be overconfident. There's no rush, so take your time. Think, and check your answer sheet. When you feel you've done your best and can do no more, return the test booklet, answer sheet and scratch pad to the examiner.

The Volunteer-Examiner Team will grade the exam right away. 74% is the passing mark. (That means no more than 6 incorrect answers.) If you are already licensed, and you pass the exam elements required to earn a higher class of license, you will receive a certificate allowing you to operate with your new privileges. The certificate has a special identifier code that must be used on the air when you use your new privileges, until your permanent license arrives from the FCC.

AND NOW, LET'S BEGIN

Subelement 3AA in Chapter 10 and subelement 3BA in Chapter 11 cover the rules and regulations for the Element 3A and Element 3B exams. You should use *The FCC Rule Book* to find the material covered by those questions. Then go over that section of the question pool to check your understanding of the rules. Perhaps you will want to study the rules a few at a time, using that as a break from your study in the rest of this book.

The remainder of this book will provide the background in electronics theory that you will need to pass the Element 3A Technician or Element 3B General class written exam.

Key Words

Amateur Auxiliary—a voluntary organization, administered by ARRL Section Managers. The primary objectives are to foster amateur self-regulation and compliance with the rules.

AMTOR—Amateur Teleprinting Over Radio. AMTOR provides error-correcting capabilities. See Automatic Repeat Request and Forward Error Correction.

ASCII—American National Standard Code for Information Interchange, a seven-bit digital code used in computer and radioteleprinter applications

Automatic Repeat Request (ARQ)—one of two AMTOR communications modes. In ARQ, also called Mode A, the two stations are constantly confirming each other's transmissions. If information is lost, it is repeated until the receiving station confirms reception.

Azimuth-equidistant projection map—a map made with its center at one geographic location and the rest of the continents projected from that point. Also called a great-circle map, this map is the most useful type for determining where to point a directional antenna to communicate with a specific location.

Band plan—an agreement for operating within a certain portion of the radio spectrum. Band plans set aside certain frequencies for each different mode of amateur operation, such as CW, SSB, FM, repeaters and simplex.

Baudot—a five-bit digital code used in teleprinter applications

Clipping—occurs when the peaks of a voice waveform are cut off in a transmitter because of overmodulation. Also called *"flattopping."*

Computer-Based Message Systems (CBMSs)—a system in which a computer is used to store messages for later retrieval. Also called a RTTY mailbox.

Coordinated Universal Time (UTC)—a system of standardized time, based on time at the Prime Meridian, which passes through Greenwich, England.

Forward Error Correction (FEC)—one of two modes of AMTOR communications. In the FEC mode, each character is sent twice. The receiving station checks for errors in the mark/space ratio. If an error is detected, a space is printed to show that an incorrect character was received. Also called Mode B.

Frequency coordinator—a volunteer who keeps records of repeater input and output frequencies and recommends frequencies to amateurs who wish to put new repeaters on the air.

Full-break-in (QSK)—with QSK, an amateur can hear signals between code characters. This allows another amateur to break into the communication without waiting for the transmitting station to finish.

Great-circle path—either one of two direct paths between two points on the surface of the earth. One of the great-circle paths is the shortest distance between those two points. Great-circle paths can be visualized if you think of a globe with a rubber band stretched around it connecting the two points.

International Telecommunication Union (ITU)—the international body that regulates the use of the radio spectrum

Long-path communication—communication made by pointing beam antennas in the directions indicated by the longer great-circle path between the stations. To work each other by long-path, an amateur in Hawaii would point his antenna west and an amateur in Florida would aim east.

MAYDAY—from the French "m'aider" (help me), MAYDAY is used when calling for emergency assistance in voice modes.

Offset—the difference between a repeater's input and output frequencies. On 2 meters, for example, the offset is either plus 600 kHz or minus 600 kHz from the receive frequency.

Q-signals—three-letter groups, such as QSK and QRP, used to facilitate CW communications. See Table 2-5.

Repeater—an amateur station that receives a signal and retransmits it for greater range

RST System—the system used by amateurs for giving signal reports. "R" stands for readability, "S" for strength and "T" for tone.

Simplex—a term normally used in relation to VHF and UHF operation, simplex operation means you are receiving and transmitting on the same frequency

Short-path communication—communication made by pointing beam antennas in the direction indicated by the shorter great-circle path

SOS—a Morse code call for emergency assistance

Splatter—the term used to describe a very-wide-band signal, caused by overmodulation of a sideband transmitter

Voice-Operated Relay (VOX)—circuitry that activates the transmitter when the operator speaks into the microphone

Chapter 2

Operating Procedures

On-the-air operation is the thrill of Amateur Radio. There are many different modes of operation available, and true enjoyment of the hobby comes from taking advantage of this diversity. This chapter covers rules and procedures normally followed on the air when using some of the modes that you, as a Technician or General class licensee, are authorized.

TELEPHONY

Radiotelephony is probably the most natural communications mode of all. Although "phone" operation appears to be as simple as grabbing a microphone and talking, it isn't. There are a few rules and guidelines to keep in mind for courteous and legal on-the-air operation. Initially, make sure your transmitter is operating properly: adjust your microphone gain correctly. This adjustment can be made to an SSB (type J3E emission) transmitter while observing a monitor oscilloscope. Proper voice waveforms have rounded peaks. Peaks with flat tops indicate *clipping* (also called *flattopping*), caused by an excessive microphone gain control setting. See Figure 2-1. If you don't have a monitor 'scope, consult the owner's manual for your rig for instructions on proper operation and gain control settings.

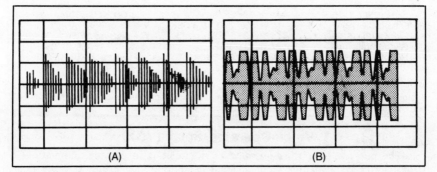

(A)   (B)

Figure 2-1—The waveform of a properly adjusted SSB transmitter is shown at A. B shows a severely clipped signal.

Generally speaking, if your rig has an automatic level control (ALC) meter, you should adjust the microphone gain until you see a slight movement of the ALC meter on modulation peaks. With an ALC indicator light, adjust the microphone gain until the indicator just lights on voice peaks.

FM transmitters (type F3E emission) should be checked with a deviation meter and adjusted for proper deviation on voice peaks. Once set, FM transmitters usually do not need to be readjusted.

Why is proper operation important? If your microphone gain is set too high, your transmitted signal will be distorted. Distortion can cause "splatter," resulting in interference to stations on other frequencies. Speak clearly into the microphone, while remaining a constant, close distance away from it. This will reduce background noise, raising the signal-to-background-noise ratio. If you use a speech processor, use the minimum necessary processing level to limit distortion. An overprocessed, distorted signal is more difficult to copy than an unprocessed one.

Speech is the primary element of radiotelephone. For effective voice communications, your speech should be clearly enunciated, delivered directly into the microphone at a speed suitable for the purpose and conditions at hand. To be efficient, the operator must use the microphone and other speech equipment properly.

Microphones vary in design and purpose, but generally it is best to speak close to the microphone, holding it a few inches away from you. This will improve the signal-to-noise ratio by making your voice louder than any background noises, such as the kids, the TV set or traffic noise when you are in your car.

In any mode, (SSB, FM and even AM) pay attention to your mic technique to keep average speech levels high without overdoing it. Automatic level controls (ALC) help, but operator care is also needed to maintain relatively constant speech levels.

Phonetics

When band conditions are poor, use phonetics when signing your call or passing important information. Standard phonetics work best, as they are universally understood. A list of standard phonetics recommended by the International Telecommunication Union (ITU) is shown in Table 2-1. For example, WA3VIL would be said phonetically as Whiskey Alfa Three Victor India Lima.

Table 2-1

International Telecommunication Union Phonetics

A—Alfa (**AL FAH**)	I—India (**IN DEE AH**)	R—Romeo (**ROW** ME OH)
B—Bravo (**BRAH** VOH)	J—Juliett (**JEW** LEE ETT)	S—Sierra (SEE **AIR** RAH)
C—Charlie (**CHAR** LEE or **SHAR** LEE)	K—Kilo (**KEY** LOH)	T—Tango (**TANG** GO)
	L—Lima (**LEE MAH**)	U—Uniform (**YOU** NEE FORM or **OO** NEE FORM)
D—Delta (**DELL** TAH)	M—Mike (**MIKE**)	
E—Echo (**ECK** OH)	N—November (NO **VEM** BER)	V—Victor (**VIK** TAH)
F—Foxtrot (**FOKS** TROT)	O—Oscar (**OSS** CAH)	W—Whiskey (**WISS** KEY)
G—Golf (GOLF)	P—Papa (PAH **PAH**)	X—X-RAY (**ECKS** RAY)
H—Hotel (HOH **TELL**)	Q—Quebec (KEH **BECK**)	Y—Yankee (**YANG** KEY)
		Z—Zulu (**ZOO** LOO)

Note: The **Boldfaced** syllables are emphasized. The pronunciations shown in this table were designed for speakers from all international languages. The pronunciations given for "Oscar" and "Victor" may seem awkward to English-speaking people in the US.

Voice-Operated Relay (VOX)

The purpose of a *voice-operated relay* circuit is to provide automatic TR (transmit/receive) switching within an amateur station. When using VOX, the transmitter switches on, the antenna is connected to the transmitter, and the receiver is muted as soon as you begin to speak; when you stop talking, the transmitter is turned off and the antenna is connected to the receiver. VOX circuits can be used with separate transmitter/receiver combinations, and most modern transceivers have a VOX circuit built in.

For CW operation, a VOX circuit can be used to key the transmitter when you begin sending, and to shut it off when you stop. With proper adjustment, you can hear other signals between words in your transmission. This type of operation is called semi-break-in CW.

Three controls must be adjusted before using VOX: VOX sensitivity, anti-VOX and VOX delay. These adjustments will vary with different pieces of equipment, and even with different operators using the same gear. Since VOX activates the transmitter, any VOX circuit control adjustments should be made with the transmitter connected to a dummy load.

Normally, VOX sensitivity is adjusted first. The setting of this control determines the level of audio input necessary to activate the transmitter. With the VOX delay and anti-VOX controls set to the middle of their adjustment ranges, turn the receiver volume to minimum. Speak into the microphone and slowly increase the setting of the VOX sensitivity control until the transmitter activates. Stop talking and allow the transmit relay to switch back to receive. Speak again, and "fine-tune" the sensitivity adjustment: your voice should activate the transmitter instantly, but background noise in the shack shouldn't.

Next, adjust the VOX delay. This part of the VOX circuit controls the amount of time the transmitter remains activated after you stop talking. Set it so the transmitter deactivates ("drops") when you stop to take a breath between thoughts. That way, a radiotelephone contact can approach a face-to-face conversation.

The anti-VOX control is adjusted last. Anti-VOX prevents receiver audio from tripping the transmitter. It accomplishes this by applying a portion of the receiver audio to the microphone amplifier circuit 180 degrees out of phase with the actual receiver audio. The anti-VOX control is used to vary the feedback level. If the anti-VOX control is set too high, you'll have trouble activating the VOX circuit when you speak into the microphone. Insufficient anti-VOX will cause receiver audio to trip the transmitter. The VOX sensitivity and anti-VOX control settings may interact slightly, so you may need to readjust the VOX sensitivity adjustment after setting the anti-VOX level.

Signal Reporting

One piece of information that is exchanged during almost every amateur QSO is a signal report. A standard system of reporting signal readability, strength and tone has been developed to describe a CW signal. The *"RST" System* is shown in Table 2-2. R stands for readability, S for strength and T for tone. This system has three scales: 1 through 5 for readability and 1 through 9 each for strength and tone.

Table 2-2

The RST System

Readability

1 — Unreadable
2 — Barely readable, occasional words distinguishable
3 — Readable with considerable difficulty
4 — Readable with practically no difficulty
5 — Perfectly readable

Signal Strength

1 — Faint signals barely perceptible
2 — Very weak signals
3 — Weak signals
4 — Fair signals
5 — Fairly good signals
6 — Good signals
7 — Moderately strong signals
8 — Strong signals
9 — Extremely strong signals

Tone

1 — 60-cycle ac or less, very rough and broad
2 — Very rough ac, very harsh and broad
3 — Rough ac tone, rectified but not filtered
4 — Rough note, some trace of filtering
5 — Filtered rectified ac but strongly ripple-modulated
6 — Filtered tone, definite trace of ripple modulation
7 — Near pure tone, trace of ripple modulation
8 — Near perfect tone, slight trace of modulation
9 — Perfect tone, no trace of ripple or modulation of any kind

---

On CW, a report of RST 368 would be interpreted as "Your signals are readable with considerable difficulty, good strength, with a slight trace of modulation." The tone report is a useful indication of transmitter performance. When the RST system was developed the tone of amateur transmitters varied widely; today, a tone report of 7 or less is cause to ask a few other amateurs for their opinion of the transmitted signal. Consistent poor tone reports may mean that you have problems in your transmitter.

On phone, the tone report is not used. Signal reports are given in the Readability-Strength format. A very good signal is 58 or 59. Above S9, strength reports are given as decibels (dB) above S9. A report of S9+10 indicates that your signal is 10 dB over S9.

Don't spend a lot of time worrying about what signal report to give a station you are in contact with. The scales used are rather broad, and are simply an indication of how you are receiving the other station. As you gain experience with the descriptions shown in Table 2-2, you will be more comfortable at estimating the proper signal report. You should not simply return the same report that you receive. Give an honest estimate of the other station's signals. It is entirely possible that the other station is hearing you perfectly, with extremely strong signals (59) while you are having considerable difficulty copying his or her weak signals (33).

[If you are studying for an Element 3A (Technician) exam, turn to Chapter 10 and study questions 3B-1.1 through 3B-1.3 and 3B-1.6 through 3B-1.16.

If you are studying for an Element 3B (General) exam, study questions 3B-1.4, 3B-1.5, 3B-4.1 and 3B-4.2 in Chapter 11.]

Repeater Operation

A *repeater* is an amateur station that receives a signal and retransmits (repeats) that signal. Often located on top of a tall building or a high mountain, repeaters greatly extend the operating coverage of amateurs using mobile and hand-held VHF and UHF transceivers. Most repeaters used for voice communication receive on one frequency and simultaneously retransmit the signal on another frequency. Some repeaters (particularly those designed for digital communications) hold the message for transmission later. These repeaters may receive and transmit on the same frequency.

To use a repeater, you must have a transceiver with the capability of transmitting on the repeater input frequency (the frequency the repeater receives on), and receiving on the repeater output frequency (the frequency the repeater transmits on). On a modern VHF transceiver using a frequency synthesizer you simply dial in the correct frequency. On some rigs you must set both the transmit and receive frequency; on other rigs you set only the receive frequency and set an "offset" switch to the correct position to provide the desired transmit frequency. The offset switch controls the separation between receive and transmit frequencies. Older rigs may require the installation of new crystals for each desired operating frequency. Consult your owner's manual for complete instructions.

With an "open" repeater, all you have to do is transmit to access the repeater. You will then be able to talk to anyone else within range of the repeater. Some repeaters (called "closed" repeaters) require the transmission of a subaudible tone or series of tones or bursts to gain access. This type of repeater cannot be accessed simply by keying the microphone—you must transmit the correct subaudible tone or series of tones in order to get into the repeater.

On most repeaters, the transmitting carrier will continue for a few seconds after a user stops transmitting. This is called the "squelch tail." If you don't hear the squelch tail at the end of your transmission, then you may not have accessed the repeater.

Most repeaters limit the length of a single transmission with a "time-out timer." This ensures that the repeater will turn itself off if the transmitter is on continuously, and it also prevents one person from monopolizing the repeater. Some repeaters send a short tone (some use the Morse code letter κ) to indicate that the timer has reset and the next person may transmit without letting the squelch tail drop. On other repeaters you may have to let the squelch tail drop to reset the timer. Ask a regular user of the repeater for proper operating etiquette on a particular machine.

To use repeaters properly, you should become acquainted with the operating practices that are inherent in this unique mode:

1) Monitor the repeater to become familiar with any peculiarities in its operation.

2) To initiate a contact, simply indicate that you're listening on the frequency. In different geographical areas there are different practices for making yourself heard, but generally something like, "This is KA1IXI monitoring" will suffice. It is not generally considered good operating practice to call CQ through a repeater. If you are looking for a specific person on the repeater, simply call once: "K1XA, this is N2YL," for example.

3) Identify legally; you must transmit your call sign at the end of each contact and every 10 minutes in between.

4) Pause between transmissions. This allows other hams to use the repeater (someone may have an emergency).

5) Keep transmissions short and thoughtful. Your monologue may prevent someone from using the repeater in an emergency situation.

6) Use *simplex* whenever possible. If you can complete your QSO on a direct frequency, there's no need to tie up the repeater and prevent others from using it.

7) Use the minimum amount of power necessary to maintain communications. Not only does Section 97.67 of the FCC rules require this, but it minimizes the possibility of accessing distant repeaters on the same frequency.

8) Don't break into a contact unless you have something to add. If you wish to join a conversation in progress, wait for a pause in the conversation and transmit your call sign once. Use the phrase "break" or "break break" to interrupt a QSO *only* if you have emergency traffic.

9) FCC forbids using autopatch (facilities that connect the repeater to the telephone system) for anything that could be construed as business communication. Nor should an autopatch be used to avoid a toll call. Do not use an auto-patch where regular telephone service is available.

10) All repeaters are assembled and maintained at considerable expense and effort. Usually, an individual or group is responsible and it is proper that regular users of a machine support the effort of keeping it on the air.

The Repeater Directory

The ARRL publishes a *Repeater Directory*, a listing of over 8000 repeaters in the US and Canada. If you own a rig that operates above 29.5 MHz, you'll find the *Repeater Directory* useful for locating repeaters in your area and indispensable when you're traveling away from home.

Frequency Coordination

Frequency coordinators are volunteers (not appointed by ARRL) who keep extensive records of repeater input, output and control frequencies, including those repeaters not listed in directories (sometimes at the owner's request). The coordinator will recommend frequencies for a proposed repeater in order to minimize interference with other repeaters and simplex frequencies. When a conflict arises, FCC acknowledges the rights of a coordinated repeater over those of another repeater placed into operation without obtaining a coordinated frequency. The Commission has taken action against noncoordinated repeaters that caused interference to coordinated repeaters. Anyone considering the installation of a repeater should check with the local frequency coordinator. For the name and address of the frequency coordinator in your area, check with the latest edition of the *Repeater Directory* or contact ARRL Headquarters.

Band Plans

Band plan refers to an agreement among concerned radio amateurs for operating within a certain portion of the radio spectrum. The goal is to minimize interference among the various modes sharing each band by setting aside certain sections of a band for each different operating mode. Besides FM repeater and simplex activity, you can find CW, SSB, AM, satellite and radio-control operations (among others) on the VHF bands. On the amateur bands between 3.5 and 148 MHz, FCC-mandated subband restrictions apply, but different operating modes must still share spectrum space (RTTY and CW operators, and slow-scan television and phone operators, for example). Voluntary band plans help minimize conflicts. The ARRL 144- to 148-MHz band plan is shown in Table 2-3. More information on VHF and UHF band plans can be found in the ARRL *Repeater Directory*; HF band plans are covered in *The ARRL Operating Manual*.

Table 2-3

144-148 MHz

The following band plan has been proposed by the ARRL VHF-UHF Advisory Committee.

144.00-144.05	EME (CW)
144.05-144.06	Propagation beacons
144.06-144.10	General CW and weak signals
144.10-144.20	EME and weak signal SSB
144.20	National calling frequency
144.20-144.30	General SSB operation
144.30-144.50	New OSCAR subband
144.50-144.60	Linear translator inputs
144.50-144.90	FM repeater inputs
144.90-145.10	Weak signal and FM simplex
145.10-145.20	Linear translator outputs
145.10-145.50	FM repeater outputs
145.50-145.80	Miscellaneous and experimental modes
145.80-146.00	OSCAR subband
146.01-146.37	Repeater inputs
146.40-146.58	Simplex
146.61-147.39	Repeater outputs
147.42-147.57	Simplex
147.60-147.99	Repeater inputs

Input/Output Separation

One function of band plans is to specify certain pairs of frequencies for repeater inputs and outputs. The separation between inputs and outputs (called the *offset*) varies from band to band—on 10 meters, the separation is 100 kHz. The N8EEG repeater in Athens, Ohio uses an input of 29.540 MHz with a 29.640 MHz output. On 6 meters, the split is an even 1 megahertz. The K1FFK repeater on Mount Greylock in Massachusetts has its input on 52.23 MHz and its output on 53.23 MHz.

On 2 meters, the offset is 600 kHz. The W9MQD repeater in Slinger, Wisconsin has its input on 146.13 MHz and its output on 146.73 MHz. Between 144 and 147 MHz, the input of a repeater is usually on the lower frequency and the output on the higher. Between 147 and 148 MHz, the input is usually higher than the output frequency.

On the 220-MHz band (1.25 meters), the separation is 1.6 MHz—you'll find the input of the W1AW 220 machine at 223.24 MHz and the output at 224.84 MHz. Repeaters on the 450-MHz band (70 centimeters) use a 5-MHz split. KØTS in Rochester, Minnesota has its input on 449.80 MHz and its output on 444.80 MHz.

[If you are studying for an Element 3A (Technician) exam, turn to Chapter 10 and study questions 3B-3.1 through 3B-3.7, 3B-3.9 through 3B-3.11 and 3B-13.

If you are studying for an Element 3B (General) exam, study questions 3B-3.8 and 3B-3.12 in Chapter 11.]

DIGITAL COMMUNICATIONS

Amateur Radio digital communications normally involve the use of the *ASCII*, *Baudot* or *AMTOR* codes and the original digital code, Morse. The Baudot code, also known as the International Telegraph Alphabet Number 2 (or ITA2) was the only digital code (other than Morse) allowed on the amateur bands until 1980. The Baudot code uses five information bits, with additional bits (called "start" and "stop" bits) to indicate the beginning and end of the character. ASCII stands for American National Standard Code for Information Interchange; ASCII is the commonly used code for computer systems. The ASCII code uses seven information bits, so more characters can be defined than with the five-bit Baudot code. AMTOR, which stands for Amateur Teleprinting Over Radio, combines a modified Baudot code with error-correcting capabilities.

As with radiotelephone operation, there are a few standard operating practices for the digital modes. We'll cover a few of the main points here. For more details, consult *The ARRL Operating Manual.*

Radioteletype

Radioteleprinter operation uses *frequency-shift keying (FSK)* to convey information. This shift is between two frequencies, called "mark" and "space." On HF, where frequency-shift keying (FSK), or mode F1B is used, the mark and space are normally 170 Hz apart. On VHF, where audio-frequency shift keying (AFSK), or F2B, is used, a variety of shifts between mark and space frequencies are used. Some common shifts are 170 Hz, 850 Hz and 1 kHz. Presently, the most common shift is 170 Hz. Common speeds for HF radioteletype (RTTY) operation are 60, 75 and 100 WPM (45.45, 56 and 75 bauds).

Table 2-4

Suggested US RTTY Operating Frequencies (kHz)

3590 RTTY DX
3610-3630
7040 RTTY DX
7090-7100
10,140-10,150
14,075-14,100
18,100-18,110*
21,090-21,100
24,920-24,930
28,090-28,100

*Pending FCC approval for US amateur use.

On the HF bands, most of the RTTY operation is found at the top of the CW portion of each band. Traditionally, the 20-meter RTTY subband has received the heaviest use during daylight hours; the 80-meter RTTY subband is active at night. Recently, the increased popularity of RTTY has brought about greater use of the 40-meter RTTY subband. Suggested operating frequencies for RTTY are shown in Table 2-4.

To call CQ on RTTY, you might use this standard calling sequence:

CQ CQ CQ DE W1AW W1AW W1AW
CQ CQ CQ DE W1AW W1AW W1AW
CQ CQ CQ DE W1AW W1AW W1AW

THIS IS CHUCK IN NEWINGTON CT

K

If you copied this CQ and you want to answer the station, here's the accepted method:

W1AW W1AW W1AW DE N1JT N1JT N1JT K

RTTY operators will occasionally send a string of the letters "R" and "Y": RYRYRYRYRY. The letter "R" in the Baudot code has a space for bits 1, 3 and

5 and a mark for the other two bits; "Y" is just the opposite, with a space in bits 2 and 4 and mark in bits 1, 3 and 5. A string of "RY"'s alternates between the mark and space bits, and is useful for testing equipment. This also helps other amateurs tune their equipment.

If you want to break into a RTTY conversation, wait until one station finishes transmitting. Then quickly send "BK DE (your call)." If you're acknowledged, make a short transmission and turn it back to the next station in the group.

Computer-Based Message Systems (RTTY Mailboxes)

Since the mid-1970s, a number of *Computer-Based Message Systems (CBMSs)* have appeared on the ham bands. You may have heard them called Message Storage Operations (MSOs), bulletin boards or mailboxes. In a CBMS, a computer is used to store messages for later retrieval.

Despite their variations, these systems have some features in common. They will automatically respond to calls on their operating frequency if the calling station uses the correct character sequence. A remote station can send messages to be stored in the system, and a station may request a listing of the messages on file. Stations can read remotely entered messages plus any bulletin messages. Some messages may be password protected to restrict access.

Message handling in this manner is third-party traffic, and the system operator (SYSOP) is required to observe appropriate rules concerning message content. The SYSOP is also responsible for maintaining control of the transmitter and taking it off the air in the event of a malfunction.

AMTOR

The benefit of using AMTOR is its error detection and correction properties. There are two AMTOR modes: Mode A, called *Automatic Repeat Request (ARQ)*, and Mode B, called *Forward Error Correction (FEC)*.

When two stations are in Mode A AMTOR communication, they are constantly confirming each other's transmissions: If a piece of information is missed or lost (because of interference or any other reason), it is repeated until the receiving station confirms reception. The transmitting station sends three characters, then pauses for the receiving station to confirm or deny correct reception. The same block of three characters is repeated or the next three are sent, depending on the reply. This interactive "handshaking" takes 450 milliseconds for every cycle between the two stations.

Mode B AMTOR is a "broadcast" mode, in which the transmitting station sends each character twice. The receiving station examines each character as it is received, looking for any errors in the mark/space ratio (each character contains four mark elements and three space elements). The first character of the pair is stored in memory as it is received. When the second character is received, the character pair is checked for errors. If one or both characters contain the proper mark/space ratio, the character is printed on the screen. If neither character has the correct ratio, the receiving station records an error signal, signified by a space.

To establish an AMTOR contact, get on a calling frequency and make a general call in Mode B. (Common calling frequencies are 14.075 MHz and 3.6375 MHz.) Once contact is established, switch to Mode A.

ASCII

ASCII is a coded character set used for computer systems and related equipment. ASCII differs from the Baudot code in that it has a larger character set. The 7 information bits in each character make it possible for the ASCII character set to provide upper- and lower-case letters, numbers, punctuation and special characters.

To establish a contact using ASCII, use the same procedure described for Baudot RTTY.

HF ASCII transmissions are normally sent at 110 or 300 bauds, using 170-Hz shift. Higher data rates are used on higher-frequency bands. On 28 MHz and above, for instance, the FCC permits use of 1200 bauds. As the speed of an RTTY transmission is increased, the bandwidth required for that transmission also increases. More spectrum space is available on the VHF and UHF bands, so the FCC permits the use of speeds up to 19.6 kilobauds on frequencies above 50 MHz and up to 56 kilobauds above 220 MHz. More about the technical requirements and restrictions on RTTY will be covered in Chapter 8.

ASCII is also used for packet-radio communications. Packet radio is a communications mode that allows transmission of short bursts (packets) of information at high speed. The format (protocol) of information packets allows several stations to hold conversations on the same frequency at the same time without interfering with each other. Packet-radio transmissions are sent at 1200 bauds on the 2-meter band, and some very-high-speed links (9600 bauds) have been set up on 220 MHz. For more information on packet radio, see Chapter 19 of *The ARRL Handbook for the Radio Amateur*.

[If you are studying for an Element 3A (Technician) exam, turn to Chapter 10 and study question 3B-2.5 and questions 3B-2.7 through 3B-2.9.

If you are studying for an Element 3B (General) exam, study questions 3B-2.1 through 3B-2.4, question 3B-2.6 and questions 3B-2.10 through 2.12 in Chapter 11.]

CW Operating Procedures

Take some time and tune around the bands. Hear that station whose sending (fist) is so poor that the letters run together and you can't copy anything? How about the one that sends CQ at 25 WPM, only to slow down to 10 WPM for the QSO? Or how about the one that sends dots at 20 WPM and dashes at 10 WPM? Or how about the one ... You get the idea.

To sound like a pro, take a bit of time and practice good CW procedures. Why not tape record your sending and then play it back to get an idea of what you sound like to someone else? Work on the areas that need improvement. A good fist and skilled operating techniques help make you popular on the air.

Here are some items to keep in mind:

1) Listen before asking if a frequency is busy.

2) Send QRL? ("Is this frequency in use?") before calling CQ and then listen for an answer.

3) Don't send faster than you can copy.

4) Send short CQs and listen between each, rather than sending one l-o-n-g CQ. No one wants to talk with you so much that they will endure a 2, 3 or even 5-minute CQ.

5) Don't over-identify during a QSO. Send your call a few times at the beginning—some operators need several chances to copy it accurately. But after that, you need only ID every 10 minutes. Turn the transmission back to the other station by sending BK. That's much less tedious and more conversational, and makes CW operation fun.

6) Be courteous. After you tune around the bands a bit, it may seem that this old-fashioned virtue has gone out of style. But this is not necessarily so. You'll get a lot further in the long run if you are always courteous. Setting a good example is an effective way of influencing your fellow hams. Courtesy pays—it makes the bands more pleasant for all of us.

7) Learn and use procedural signals (prosigns) and Q-signals correctly.

In order for amateurs to understand each other, we must attempt to standardize our communications. You'll find, for instance, that the use of abbreviations on CW is widespread: It only makes sense—it's much easier to send a couple of letters than

Table 2-5

Abbreviations for CW Work

The purpose of using abbreviations is to reduce the amount of time required to transmit intelligence. However, many CW amateur abbreviations have been carried over into voice work and have become part of the "lingo." In general, it is considered good practice to abbreviate on CW, poor practice to use CW abbreviations on phone. The list below covers only those abbreviations most used.

AA	All after	NIL	Nothing; I have nothing for you
AB	All before		
ABT	About	NR	Number
ADR	Address	NW	Now; I resume transmission
AGN	Again	OB	Old boy
ASCII	American National Standard Code for Information interchange	OM	Old man
		OP/OPR	Operator
		OT	Old timer; old top
BCI	Broadcast interference	PBL	Preamble
BCNU	Be seeing you	PSE	Please
BK	Break; break me; break in	R	Received solid
BN	All between; been	RCVR	Receiver
BUG	Semi-automatic key	RECD/RCVD	Received
B4	Before	RFI	Radio frequency interference
C	Yes	RPT	Repeat; I repeat
CFM	Confirm; I confirm	RTTY	Radioteletype
CK	Check	SASE	Self-addressed, stamped envelope
CL	I am closing my station; call		
CLD/CLG	Called; calling	SIG	Signature, signal
CQ	Calling any station	SKED	Schedule
CUL	See you later	SRI	Sorry
CW	Continuous wave (i.e., radiotelegraph)	SSB	Single sideband
		SVC	Service; prefix to service message
DLVD	Delivered		
DR	Dear	SWL	Short wave listener
DX	Distance	TFC	Traffic
ES	And	TMW	Tomorrow
FB	Fine business; excellent	TNX/TKS	Thanks
FREQ	Frequency	TT	That
GA	Go ahead, good afternoon	TU	Thank you
GB	Good-bye	TVI	Television interference
GBA	Give better address	UR/URS	Your; you're; yours
GE	Good evening	VFO	Variable frequency oscillator
GG	Going	VY	Very
GM	Good morning	WA	Word after
GN	Good night	WB	Word before
GND	Ground	WRD	Word; words
HI	The telegraphic laugh; high	WKD/WKG	Worked; working
HR	Here; hear	WL	Will
HV	Have	WPM	Words per minute
HW	How	WX	Weather
ITU	International Telecommunication Union	XMTR	Transmitter
		XTAL	Crystal
LID	A poor operator	XYL (YF)	Wife
MSG	Message; prefix to radiogram	YL	Young Lady
N	No	Z	Coordinated Universal Time
NBVM	Narrow band voice modulation	ZB	Zero beat
NCS	Net Control Station	73	Best regards
ND	Nothing doing	88	Love and kisses

it is to spell out one or several words. There's no point in using abbreviations, however, if no one else understands what you're trying to say! Over the years, some "standard" abbreviations have been developed (see Table 2-5). In addition to these abbreviations, amateurs use procedural signals (prosigns) and Q-signals to aid in establishing and carrying on a QSO. A list of some common Q-signals is given in Table 2-6. Communications prosigns can be found in Table 2-7.

Table 2-6

Q Signals

Quite a number of International Q signals are widely used in amateur CW work, and to some extent are also used in voice transmissions—often incorrectly. The meanings of some Q signals change from time to time as dictated by the International Telecommunication Union, but most of them remain the same year after year. Nevertheless, amateurs should strive to keep up to date in the meanings of the Q signals. Listed below are most of those having a useful amateur meaning. The Q abbreviations take the form of questions only when followed by a question mark.

QRA	What is the name of your station?	QSG	Shall I send...messages at a time?
QRG	What's my exact frequency?	QSK	Can you work break in?
QRH	Does my frequency vary?	QSL	Can you acknowledge receipt?
QRK	What is my signal intelligibility? (1-5)	QSM	Shall I repeat the last message sent?
QRL	Are you busy?	QSO	Can you communicate with...direct?
QRM	Is my transmission being interfered with?	QSP	Will you relay to...?
QRN	Are you troubled by static?	QSV	Shall I send a series of Vs?
QRO	Shall I increase transmitter power?	QSW	Will you transmit on...?
QRP	Shall I decrease transmitter power?	QSX	Will you listen for...on...?
QRQ	Shall I send faster?	QSY	Shall I change frequency?
QRS	Shall I send more slowly?	QSZ	Shall I send each word/group more than once? (Answer, send twice or...)
QRT	Shall I stop sending?		
QRU	Have you anything for me? (Answer in negative.)	QTA	Shall I cancel message number...?
QRV	Are you ready?	QTB	Do you agree with my word count? (Answer negative.)
QRW	Shall I inform...that you are calling him on...?	QTC	How many messages have you to send?
QRX	When will you call again?	QTH	What is your location?
QRZ	Who is calling me?	QTR	What is the correct time?
QSA	What is my signal strength? (1-5)	QTX	Will you keep your station open for further communication with me?
QSB	Are my signals fading?		
QSD	Is my keying defective?	QUA	Have you news of...?

Table 2-7

Communications Procedures

Voice	Code	Situation
go ahead	K	used after calling CQ, or at the end of a transmission, to indicate any station is invited to transmit.
over	$\overline{AR}$	used after a call to a specific station, before the contact has been established.
---	$\overline{KN}$	used at the end of any transmission when only the specific station contacted is invited to answer.
stand by or wait	$\overline{AS}$	a temporary interruption of the contact.
roger	R	indicates a transmission has been received correctly and in full.
clear	$\overline{SK}$	end of contact. $\overline{SK}$ is sent before the final identification.
leaving the air or closing station	CL	indicates that a station is going off the air and will not listen for or answer any further calls. CL is sent after the final identification.

QSK Operation

Wouldn't it be advantageous to be able to hear what's happening on frequency while you're transmitting? *Full-break-in (QSK)* operation on CW allows this: You can listen to your transmit frequency between character elements (dots and dashes) and between words.

QSK operation is especially useful when you are sending message traffic—the receiving station can send a few dots and "break" (stop) the transmitting station for repeats (fills) of missed words. The transmitting station can also hear if another station comes on frequency and interferes with the QSO.

Some transceivers use the VOX circuitry to produce a semi-break-in operation. With this type of circuit the VOX delay is set to hold the transmitter on during the transmission of entire characters. But the receiver can come back on between characters, so at speeds above about 15 WPM the effect is almost the same as full break-in. Semi break-in is not QSK, however.

[If you are studying for an Element 3B (General) exam, study questions 3B-5.1 and 3B-5.2 in Chapter 11.]

GENERAL OPERATING PRACTICES

Operating Courtesy

On today's crowded ham bands, operating courtesy becomes extremely important. Operating courtesy is really nothing more than common sense; a good operator sets an example for others to follow. A courteous amateur operator always listens before transmitting so as not to interfere with stations already using the frequency. If the frequency is occupied, move to another frequency far enough away to prevent interference (QRM). Know how much bandwidth each mode occupies and act accordingly. A good rule of thumb is to leave 150 to 500 Hz between CW (A1A emission) signals, at least 3 kHz between suppressed carriers in single sideband (J3E) and 250 to 500 Hz center-to-center for RTTY (F1B) signals.

A good amateur operator transmits only what is necessary to accomplish his or her purpose and keeps the contents of the transmission within the bounds of propriety and good taste. Also, a good operator always reduces power to the minimum necessary to carry out the desired communications. This is not only operating courtesy, it's an FCC rule! Section 97.67[b] states "...amateur stations shall use the minimum transmitting power necessary to carry out the desired communications." A complete copy of Part 97 appears in *The FCC Rule Book*, published by the ARRL.

It shows inexcusably bad manners to interfere with a station thousands of miles away because you are talking to a buddy across town on a DX band. Judicious selection of a band is important to reduce unintentional QRM. Use VHF and UHF frequencies for local communications.

If a conflict arises between operators using simplex and a repeater on the VHF bands, it's common courtesy for the simplex operators to change frequency. It is a simple matter for the simplex operators to find a new frequency, but it is impractical to change the operating frequency of a repeater. During commuter rush hours, mobile operators should have the highest priority on the repeater (with the exception of emergency traffic, of course). All other users, such as third-party traffic nets, should relinquish the repeater to the mobile operators. Repeater operating courtesy also includes keeping your transmissions short and to the point. A long-winded monologue will make the repeater shut off (or "time out"). The repeater will usually reset itself a short time later, but timing out a repeater can be embarrassing.

Testing and "loading up" a transmitter should be done into a dummy antenna (or dummy load) without putting a signal on the air. On-the-air tests should be made only when necessary and kept as brief as possible while conducting the test. If your amateur station is within a mile of an FCC monitoring station or you are planning portable operation near a monitoring station, you should consult with the staff at the monitoring station to ensure that your amateur operations do not cause them any harmful interference. Always be more courteous than the other person—you contribute to the image of the Amateur Radio Service as being a self-disciplined, self-policing service with high standards.

[If you are studying for an Element 3A (Technician) exam, turn to Chapter 10 and study questions 3B-6.4 through 3B-6.10.

If you are studying for an Element 3B (General) exam, study questions 3B-6.1 through 3B-6.3 in Chapter 11.]

International Communications

One of the basic purposes of the Amateur Radio Service is stated in Section 97.1 of the rules: "Continuation and extension of the amateur's unique ability to enhance international good will." The opportunity for international communications occurs every minute on the amateur bands: This is what makes Amateur Radio so special. Since our radio signals can transcend national boundaries, an organization having a wider scope than national governments is necessary.

This organization is the *International Telecommunication Union (ITU)*. The ITU regulates the use of the airwaves and sets standards to help avoid confusion and political difficulties. ITU regulations state that communications between amateurs in foreign countries must be limited to technical topics or "remarks of a personal nature of relative unimportance"—you can talk about your new rig or what you had for dinner, but don't get into a serious political discussion with a foreign ham.

Under the ITU, the world has been separated into three areas, or Regions. See Figure 2-2. Amateurs in different Regions may not be able to use the same frequencies or operating modes as US operators. The United States, including Alaska and Hawaii, is in ITU Region 2. Region 3 comprises most of the islands in the Pacific Ocean, including American Samoa, the Northern Mariana Islands, Guam and Wake Island. Also included in Region 3 are China, India and parts of the Middle East. Region 1 comprises Europe, Africa, the USSR and the remainder of the Middle East.

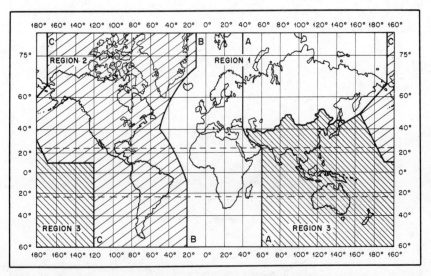

Figure 2-2—This map shows the world as divided into three ITU Regions.

Short or Long Path?

The best propagation path between any two points on the earth's surface is usually by the shortest direct route—the *great-circle path* that can be found by stretching a string tightly between the two points on a globe. If we replace the string with a rubber band going around the globe, we can see another, longer great-circle path. With the proper conditions, *long-path communication* is possible. To attempt communication by long-path propagation, you must rotate your antenna 180° from the "normal" (*short-path*) direction. For instance, using long-path propagation, an amateur in eastern Pennsylvania could work a station in Hawaii by pointing his beam a few degrees north of east—the Hawaiian amateur would point his beam a few degrees south of west. It is interesting to note that because Antarctica is at the South Pole, the short-path heading from anywhere to Antarctica is south, or 180° and the long-path heading is north, or 0°.

The long path seems to work best for communications between two places when the path is mostly over water. Since sunlight is an important ingredient for good ionospheric propagation on frequencies above 7.5 MHz, long-path communication also seems to work best when the signals will travel mostly through sunlight. From the East Coast of the United States, for example, you might try the long path to reach Hawaii just before your sunrise. If you are in Hawaii, you might try pointing your antenna to the west around sunset and shortly after that to contact stations on

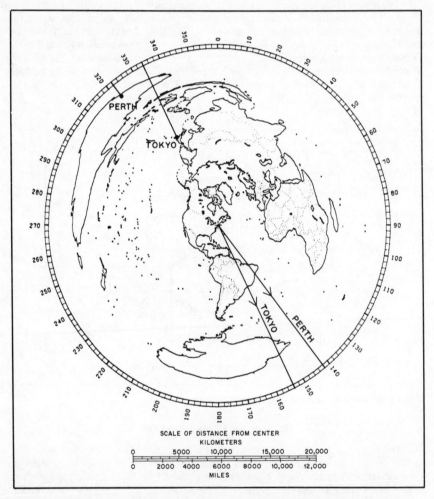

Figure 2-3—N5KR's computer-generated azimuthal-equidistant projection map centered on Newington, Connecticut. Information showing long paths to Perth and Tokyo have been added. Notice that the paths in both cases lie almost entirely over water, rather than over land masses.

the East Coast. It is important that stations on both ends of a long-path circuit point their antennas in the long-path direction.

You can understand several aspects of long-path propagation if you become accustomed to thinking of the earth as a ball. This is easy if you use a globe frequently. If you must use a flat map to find great-circle paths, an *azimuthal-equidistant projection map* (also called a great-circle map), is a useful substitute. See Figure 2-3.

Telling Time—Coordinated Universal Time (UTC)

Keeping track of time can get pretty confusing if you're talking to hams around the world. Europe, for example, is anywhere from four to ten hours ahead of us in North America. Over the years, the time at Greenwich, England has been universally recognized as the standard time in all international affairs, including ham radio. This means that wherever you are, you and your contact will have begun the QSO at the same time and date referenced to time at Greenwich, thus avoiding the confusion that would occur if we all used our local times.

Coordinated Universal Time (called UTC) is the local time at the Greenwich Meridian (0 degrees longitude) expressed in a 24-hour format. UTC is designated by the letter Z (often expressed phonetically as ZULU on phone). The easiest way to keep track of UTC is to use a separate 24-hour clock. It is also possible to modify an electric clock to show time in a 24-hour format, as shown in Figure 2-4. Many hams have a clock with a 24-hour-format digital display set to UTC.

Twenty-four-hour time lets you avoid the confusion of AM and PM. The day begins at midnight (0000 hours). Noon is 1200 hours: the afternoon hours follow from there. (You can think of it as adding 12 hours to the normal PM time—3 PM is 1500, 9:30 PM is 2130, and so on.) Keep in mind that the day also ends at midnight (2400). Midnight can be represented by either 2400 or 0000 hours—2400 corresponds to the day just ending, and 0000 the day just beginning. A time-conversion chart is shown in Table 2-8. You should become familiar with how to convert your local time to UTC, and use UTC for all of your Amateur Radio timekeeping. Be especially sure to indicate the time in UTC on any QSL cards that you send.

[If you are studying for an Element 3B (General) exam, study questions 3B-7.1 through 3B-7.5 and questions 3B-8.1 through 3B-8.9 in Chapter 11.]

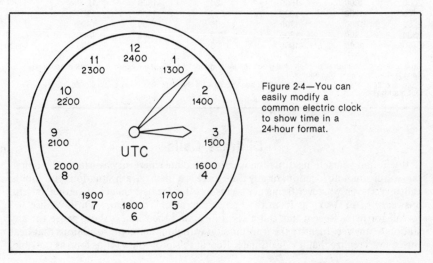

Figure 2-4—You can easily modify a common electric clock to show time in a 24-hour format.

Table 2-8

Some Fact About Time Conversion

This chart has been arranged to show UTC and the time zones used by most amateurs in North America. Coordinated Universal Time is the standard reference time used throughout the world. ARRL recommends that all amateur logging be done in UTC.

All times shown are in 24-hour time for convenience. To convert to 12-hour time: For times between 0000 and 0059, change the first two zeros to 12, insert a colon and add AM; for times between 1200 and 1259, insert a colon and add PM; for times between 1300 and 2400, subtract 12, insert a colon and add PM.

Time zone letters may be used to identify the kind of time being used. For example, UTC is designated by the letter Z, EDT/AST by the letter Q, CDT/EST by R, MDT/CST by S, PDT/MST by T, PST by U; thus, 1200R would indicate noon in the CDT/EST zone, which would convert to 1700 UTC or 1700Z.

In converting from one time to another, be sure the day or date corresponds to the new time. That is, 2100R (EST) on January 1 would be 0200Z (UTC) on January 2; similarly, 0400Z on January 2 would be 2000U (PST) on January 1.

A good method is to use UTC (Z) for all amateur logging, schedule-making, QSLing and other amateur work. Otherwise, with all the different time zones, confusion is inevitable. Leave your clock on UTC. Most of Alaska and Hawaii use W time (UTC − 10 hrs.)

UTC	EDT/AST	CDT/EST	MDT/CST	PDT/MST	PST
0000*	2000	1900	1800	1700	1600
0100	2100	2000	1900	1800	1700
0200	2200	2100	2000	1900	1800
0300	2300	2200	2100	2000	1900
0400	0000*	2300	2200	2100	2000
0500	0100	0000*	2300	2200	2100
0600	0200	0100	0000*	2300	2200
0700	0300	0200	0100	0000*	2300
0800	0400	0300	0200	0100	0000*
0900	0500	0400	0300	0200	0100
1000	0600	0500	0400	0300	0200
1100	0700	0600	0500	0400	0300
1200	0800	0700	0600	0500	0400
1300	0900	0800	0700	0600	0500
1400	1000	0900	0800	0700	0600
1500	1100	1000	0900	0800	0700
1600	1200	1100	1000	0900	0800
1700	1300	1200	1100	1000	0900
1800	1400	1300	1200	1100	1000
1900	1500	1400	1300	1200	1100
2000	1600	1500	1400	1300	1200
2100	1700	1600	1500	1400	1300
2200	1800	1700	1600	1500	1400
2300	1900	1800	1700	1600	1500
2400*	2000	1900	1800	1700	1600

Time changes one hour with each change of 15° in longitude. The five time zones in the US proper and Canada roughly follow these lines.
*0000 and 2400 are interchangeable; 2400 is associated with the date of the day ending, 0000 with the day just starting.

Distress Calls

If you find yourself in a situation where immediate emergency assistance is required (at sea, in a remote location, or any time there is a life or death situation) you should call *MAYDAY* on whatever frequency seems to offer the best chance of getting a useful answer. "MAYDAY" is from the French "m'aider" (help me). On CW, use $\overline{SOS}$ to call for help. Repeat this call a few times, and then take a short pause for any station to answer. Identify the transmission with your call sign, so stations that hear the call will realize that it is legitimate. Repeat this procedure for as long as possible, until you receive an answer. Be prepared to supply the following information to the stations that respond to an $\overline{SOS}$ or MAYDAY:

- **The location of the emergency**, with enough detail to permit rescuers to locate it without difficulty.
- **The nature of the distress.**
- **The type of assistance required** (medical aid, evacuation, food, clothing or other assistance).
- **Any other information** that might be helpful in locating the emergency area or in sending assistance.

Volunteer Monitoring

An important aspect of the Communications Amendments Act of 1982, commonly known as Public Law 97-259, is one that authorized the FCC to formally enlist the use of amateur volunteers for monitoring the airwaves for rules violations. ARRL has taken a leadership role in the volunteer monitoring function, through the Amateur Auxiliary to the FCC's Field Operations Bureau. The primary objectives of the Amateur Auxiliary are to foster amateur self regulation and compliance with the rules.

The Amateur Auxiliary is concerned with both maintenance monitoring and amateur-to-amateur interference. Maintenance monitoring is a term that refers to a program of formally enlisted amateurs monitoring the transmissions from other amateur stations for compliance with the FCC rules and technical standards. Maintenance monitoring is conducted through the ARRL Official Observer program, while amateur-to-amateur interference is handled by specifically authorized interference committees.

[If you are studying for an Element 3A (Technician) exam, turn to Chapter 10 and study questions 3B-9.1 and 3B-9.2.

If you are studying for an Element 3B (General) exam, study questions 3B-10.1 and 3B-10.2 in Chapter 11.]

Key Words

Backscatter—a small amount of signal that is reflected from the earth's surface after traveling through the ionosphere. The reflected signals may go back into the ionosphere along several paths and be refracted to earth again. Backscatter can help provide communications into a station's skip zone.

Critical angle—if radio waves leave an antenna at an angle greater than the critical angle for that frequency, they will pass through the ionosphere instead of returning to earth

Critical frequency—the highest frequency at which a vertically incident radio wave will return from the ionosphere. Above the critical frequency, radio signals pass through the ionosphere instead of returning to the earth.

D layer—the lowest layer of the ionosphere. The D layer contributes very little to short-wave radio propagation, acting mainly to absorb energy from radio waves as they pass through it. This absorption has a significant effect on signals below about 7.5 MHz during daylight.

Duct—a radio waveguide formed when a temperature inversion traps radio waves within a restricted layer of the atmosphere

E layer—the second lowest ionospheric layer, the E layer exists only during the day, and under certain conditions may refract radio waves enough to return them to earth

F layer—a combination of the two highest ionospheric layers, the F1 and F2 layers. The F layer refracts radio waves and returns them to earth. The height of the F layer varies greatly depending on the time of day, season of the year and amount of sunspot activity.

Geomagnetic disturbance—a dramatic change in the earth's magnetic field that occurs over a short period of time

Guided propagation—radio propagation by means of ducts

Ionosphere—a region in the atmosphere about 100 to 250 miles above the earth. The ionosphere is made up of charged particles, or ions.

Maximum usable frequency (MUF)—the highest frequency that allows a radio wave to reach a desired destination

Propagation—the means by which radio waves travel from one place to another

Radio-path horizon—the point where radio waves are returned by tropospheric bending. The radio-path horizon is 15 percent farther away than the geometric horizon.

Refract—to bend. Electromagnetic energy is refracted when it passes through a boundary between different types of material. Light is refracted as it travels from air into water or from water into air.

Skip zone—a region between the farthest reach of ground-wave communications and the closest range of skip propagation

Sky waves—radio waves that travel from an antenna upward to the ionosphere, where they either pass through the ionosphere into space or are refracted back to earth

Solar flux index—a measure of solar activity. The solar flux index is a measure of the radio noise on 2800 MHz.

Sudden Ionospheric Disturbance—a blackout of HF sky-wave communications that occurs after a solar flare

Sunspots—dark blotches that appear on the surface of the sun

Temperature inversion—a condition in the atmosphere in which a region of cool air is trapped beneath warmer air

Troposphere—the region in the earth's atmosphere just above the surface of the earth and below the ionosphere

Tropospheric bending—when radio waves are bent in the troposphere, they return to earth approximately 33 percent farther away than the geometric horizon

True or Geometric horizon—the most distant point one can see by line of sight

Virtual height—the height that radio waves appear to be reflected from when they are returned to earth by refraction in the ionosphere

Chapter 3

Radio-Wave Propagation

R adio amateurs have been interested in the study of radio *propagation* since the early days of radio. Amateurs have sought to understand how different factors affect radio waves as they travel from one place to another. The height of your antenna above ground, the type of antenna used, the frequency of operation, the terrain, the weather and the height and density of the *ionosphere* (a region of charged particles high above the surface of the earth) all affect how radio waves travel. It is important to understand the nature of radio waves and how their behavior is affected by the medium they travel in.

IONOSPHERIC PROPAGATION

Nearly all amateur communication on frequencies below 30 MHz is by means of *sky waves*. After leaving the transmitting antenna, this type of wave travels from the earth's surface at an angle that would send it out into space if its path were not bent enough to bring it back to earth. As the radio wave travels outward from the earth, it encounters a region of ionized particles in the atmosphere. This region, called the ionosphere, begins about 30 miles from the surface of the earth, and extends to about 260 miles. See Figure 3-1. The ionosphere refracts (or bends) radio waves, and

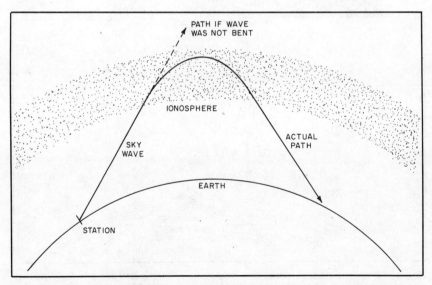

Figure 3-1—Because radio waves are bent in the ionosphere, they return to earth far from their origin. Without refraction in the ionosphere, radio waves would pass into space.

at some frequencies the radio waves are refracted enough so they return to earth at a point far from the originating station.

The earth's upper atmosphere is composed mainly of oxygen and nitrogen with traces of hydrogen, helium and several other gases. The atoms that make up these gases are electrically neutral—they have no charge and exhibit no electrical force outside their own structure. When the gas atoms absorb ultraviolet radiation from the sun, however, electrons are knocked free and the atoms become positively charged. These positively charged atoms are called ions, and the process by which they are formed is called ionization. Several ionized layers are created at different heights in the atmosphere. Each layer has a central region where the ionization is greatest. The intensity of the ionization decreases above and below this central region in each layer.

The Ionosphere: A Closer Look

The ionosphere consists of many layers of charged particles. These layers have been given letter designations, as shown in Figure 3-2.

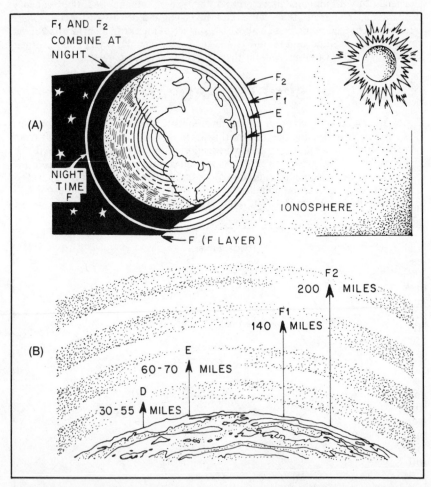

Figure 3-2—The ionosphere consists of several layers of ionized particles at different heights above the earth. At night, the D and E layers disappear and the F1 and F2 layers combine to form a single F layer.

The D Layer: The lowest layer in the ionosphere that affects propagation is called the D layer. This layer is in a relatively dense part of the atmosphere about 30 to 55 miles above the earth. The ions formed when sunlight is absorbed by the atmosphere are very short-lived in this layer—they quickly revert to their neutral atomic form. The amount of ionization in this layer varies widely depending on how much sunlight hits the layer. At noon, D-layer ionization is maximum or very close to it. By sunset, this ionization disappears.

The D layer is actually ineffective in bending high-frequency signals back to earth. The major effect the D layer has on long-distance communication is to absorb energy from radio waves. As radio waves pass through the ionosphere, they give up energy and set some of the ionized particles into motion. Lower frequencies are affected more by absorption than higher frequencies. Absorption also increases proportionally with the amount of ionization; the more ionization, the more energy the radio waves will lose as they pass through the ionosphere. Absorption is most pronounced at midday, and it is responsible for the short daytime communications ranges on the lower amateur frequencies (160, 80 and 40 meters).

The next layer of the ionosphere is the *E layer.* The E layer appears at an altitude of about 60 to 70 miles above the earth. At this height, the atmosphere is still dense enough so that ionization produced by sunlight does not last very long. This makes the E layer useful for bending radio waves only when it is in sunlight. Like the D layer, the E layer reaches maximum ionization around midday, and by early evening the ionization level is very low. The ionization level reaches a minimum just before sunrise, local time. Using the E layer, a radio signal can travel a maximum distance of about 1250 miles in one hop.

The *F layer:* The layer of the ionosphere most responsible for long-distance amateur communication is called the F layer. This layer is actually a very large region ranging from about 100 to 260 miles above the earth, depending on the season of the year, latitude, time of day and solar activity. Ionization reaches a maximum shortly after noon local standard time, but tapers off very gradually toward sunset. At this altitude, the ions and electrons recombine very slowly, so the F layer remains ionized throughout the night, reaching a minimum just before sunrise. After sunrise, ionization increases rapidly for the first few hours, and then increases slowly to its noontime maximum.

During the day, the F layer splits into two parts, F1 and F2, with central regions at altitudes of about 140 and 200 miles, respectively. These altitudes vary with the season of the year—at noon in the summer the F2 layer can reach an altitude of 300 miles. At night, the two layers recombine to form a single F layer slightly below the higher altitude. The F1 layer does not have much to do with long-distance communications; it tends to cause effects similar to those caused by the E layer. The F2 layer is responsible for almost all long-distance communication on the amateur HF bands. A one-hop radio transmission can travel a maximum distance of about 2500 miles using the F2 layer.

Virtual Height

An ionospheric layer is a region of considerable depth, but for practical purposes it is convenient to think of each layer as having a definite height. The height from which a simple reflection from the layer would give the same effects (observed from the ground) as the effects of the gradual bending that actually takes place is called the *virtual height* of the layer. See Figure 3-3.

The virtual height of an ionospheric layer for various frequencies is determined with a variable-frequency sounding device (called an ionosonde) that directs energy vertically and measures the time required for the round-trip path. As the frequency is increased, a point is reached where no energy will be returned from the ionosphere. The frequency where radio waves will penetrate a layer of the the ionosphere instead of returning to earth is called the *critical frequency* for that layer.

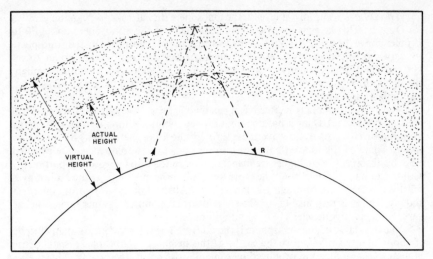

Figure 3-3—The virtual height of a layer in the ionosphere is the height at which a simple reflection would return the wave to the same point as the gradual bending that actually takes place.

Radiation Angle and Skip Distance

The lower the angle above the horizon at which a radio wave leaves the antenna, the less refraction in the ionosphere that is required to bring the wave back to earth. This is the basis for the emphasis on low radiation angles in the pursuit of DX on the HF and VHF bands.

Some of the effects of radiation angle are illustrated in Figure 3-4. The high-angle waves are bent only slightly in the ionosphere, and so pass through it. The wave at the somewhat lower angle is just able to be returned from the ionosphere. In daylight it might be returned from the E layer. The point on earth where this wave is returned from the ionosphere (A) is closer to the transmitting station than the point of return for the lowest-angle wave, which returns to B.

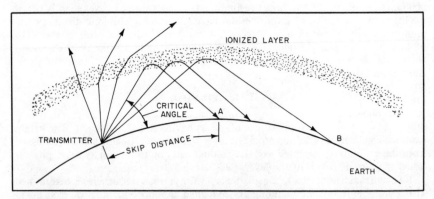

Figure 3-4—Ionospheric propagation. The waves at left leave the transmitter above the critical angle—they are refracted in the ionosphere, but not enough to return to earth. The wave at the critical angle will return to earth. The lowest-angle wave will return to earth farther away than the wave at the critical angle. This explains the emphasis on low radiation angles for DX work.

The highest radiation angle that will return a radio wave to earth under specific ionospheric conditions is called the critical angle. Waves that meet the ionosphere at greater than the *critical angle* will pass through the ionosphere into space.

Maximum Usable Frequency

There is little doubt that the critical frequency is important to amateur communication. Radio amateurs, however, are more interested in the frequency *range* over which communication can be carried via the ionosphere. What most amateurs want to know is the *maximum usable frequency* (abbreviated MUF) for a particular distance at the time of day when communication is desired. The MUF is the highest frequency that allows a radio wave to reach the desired destination, using E- or F-layer propagation. The MUF is subject to seasonal variations as well as changes throughout the day.

Why is the MUF so important? If we know the MUF, we can make an accurate prediction as to which amateur band will give us the best chance for communication via a particular path. For example, if we wish to contact a distant station and we know that the MUF for that path is 17 MHz at the time we wish to make our contact, we can predict that the closest amateur band just below that frequency (the 20-meter, or 14-MHz, band in this case) offers our best chances for a long-distance contact. There is no single MUF for a given location at any one time—the MUF will vary depending on the direction and distance to the station we wish to contact. Understanding and using the MUF is just one of the many ways we can change the study of propagation from guesswork into a science.

[If you are studying for an Element 3A (Technician) exam, turn to Chapter 10 and study the following questions: 3C-1.1 through 3C-1.5, 3C-1.8, 3C-1.11, 3C-1.12, 3C-1.14, 3C-2.1, 3C-2.2, 3C-2.5, 3C-2.6, 3C-3.1, 3C-3.2, 3C-3.5 and 3C-4.1 through 3C-4.4.

If you are studying for an Element 3B (General exam, study the following questions in Chapter 11: 3C-1.6, 3C-1.7, 3C-1.9, 3C-1.10, 3C-1.13, 3C-2.3, 3C-2.4, 3C-3.3 and 3C-3.4.]

Solar Activity and Radio-Wave Propagation

From this discussion, it should be clear that the sun plays a major role in the ionization of different layers in the ionosphere. In fact, the sun is the dominant factor in sky-wave communication. If we want to communicate over long distances, we must understand how solar conditions will eventually affect our radio signals.

You know about the day-to-day sun-related cycles on earth, such as the time of day and the season of the year. Conditions affecting radio communication vary with these same cycles. There are also long-term and short-term solar cycles that influence propagation in ways that are not so obvious. The condition of the sun at any given moment has a very large effect on long-distance radio communication, and the sun is what makes propagation prediction an inexact science.

Man's interest in the sun is older than recorded history. *Sunspots,* dark regions that appear on the surface of the sun, were observed and described thousands of years ago. Observers noted that the number of sunspots increased and decreased in cycles. The solar observatory in Zurich, Switzerland has been recording solar data on a regular basis since 1749; accordingly, the solar cycle that began in 1755 was designated cycle 1. The low sunspot numbers marking the end of cycle 20 and the beginning of cycle 21 occurred in 1976. Sunspot cycles average roughly 11 years in length, but some cycles have been as short as 9 years, and some as long as 13 years. The highs and lows of individual sunspot cycles also vary a great deal. For example, cycle 19 peaked with a smoothed mean sunspot number of over 200, yet cycle 14 peaked at only 60. When a sunspot cycle is at its peak, ionization in the atmosphere is at a maximum. High sunspot numbers, therefore, indicate good worldwide radio communications. During a peak in the sunspot cycle, the 20-meter amateur band is open to distant parts of the world almost continuously. The cycle 19 peak in 1957 and 1958 was responsible for the best propagation conditions in the history of radio. Sunspot cycles do not follow consistent patterns—there can be highs that seem to come from nowhere during periods of relatively low sunspot activity.

An important clue for anticipating variations in solar radiation levels, and radio propagation changes resulting from them, is the time it takes the sun to rotate on its axis, approximately 27 days. Active areas on the sun capable of influencing propagation may recur at four-week intervals for four or five solar rotations. If the MUF is high and propagation conditions are good for several days, you can expect similar conditions to develop approximately 27 days later.

Another useful indication of solar activity is solar flux, or radio energy coming from the sun. Increased solar activity produces higher levels of solar energy, which produces greater ionization in the ionosphere. Sophisticated receiving equipment and large antennas that can be pointed at the sun are required to measure solar flux. When this measurement is made, a number called the *solar-flux index* is given to represent the amount of solar flux. The solar-flux index is gradually replacing the sunspot number as a means of predicting radio-wave propagation.

The solar-flux measurement is taken at 1700 UTC daily in Ottawa, Canada, on 2800 MHz (10.7 centimeters); the information is then transmitted by the US Bureau of Standards station WWV in Fort Collins, Colorado. Both the sunspot number and the solar flux measurements tell us similar things about solar activity, but to get the sunspot number the sun must be visible through a telescope (with the proper filters). The solar-flux measurement may be taken under any weather conditions.

Using the Solar Flux

We can use the solar flux numbers to make general predictions about band conditions from day to day. Figure 3-5 shows a relationship between average sunspot numbers and the 2800 MHz solar-flux measurement. The solar flux varies directly with the activity on the sun, ranging from values around 60 to 250 or so. Flux values in the 60s and 70s indicate fair to poor propagation conditions on the 14-MHz (20-meter) band and higher frequencies; values from 90 to 110 or so indicate good conditions up to about 24 MHz (12 meters), and values over 120 indicate very good conditions on 28 MHz (10 meters), and even up to 50 MHz (6 meters) at times. The higher the frequency, the higher the flux value that will be required for good propagation. On the 6-meter band, for example, the flux values must be in the 200s for reliable long-range communications.

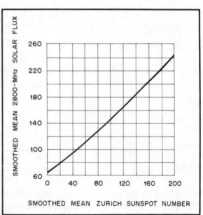

Figure 3-5—The relationship between smoothed Zurich sunspot numbers and 2800-MHz solar flux values.

Sudden Ionospheric Disturbances

One solar phenomenon that can severely disrupt HF sky-wave radio communication is a solar flare. A solar flare is a large eruption of energy and solar material from the surface of the sun. A large amount of ultraviolet radiation travels from

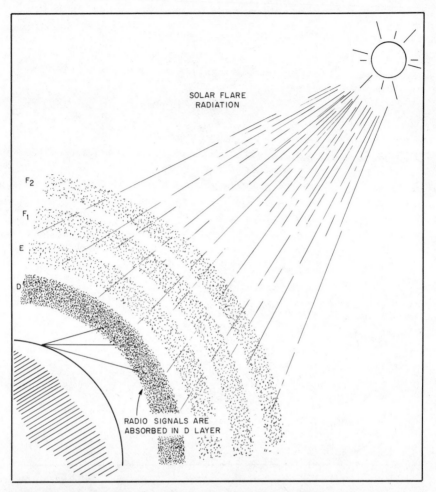

SOLAR FLARE
RADIATION

F₂

F₁

E

D

RADIO SIGNALS ARE
ABSORBED IN D LAYER

Figure 3-6—Approximately 8 minutes after a solar flare occurs on the sun, the radiation released by the flare reaches the earth. This radiation causes increased ionization and radio-wave absorption in the D layer.

the sun at the speed of light, reaching the earth approximately 8 minutes after the flare begins. When the ultraviolet radiation reaches the earth, the level of ionization in the ionosphere increases rapidly and D-layer absorption is greatly enhanced. This is called a *sudden ionospheric disturbance (SID)*. See Figure 3-6.

If the earth and the solar flare producing the radiation are directly in line, the SID will be most severe. The D layer may be so heavily ionized that HF sky-wave communications are completely blacked out. If the area of the solar flare is off to the side of the sun, the SID will be less severe. Since D-layer ionization is responsible for the disruption of HF communications during an SID, the lower frequency bands will be affected first. Communications may still be possible on a higher band. Because an SID only affects the side of the earth facing the sun, dark-path communications will be unaffected. An SID may last from a few minutes to a few hours, with conditions gradually returning to normal.

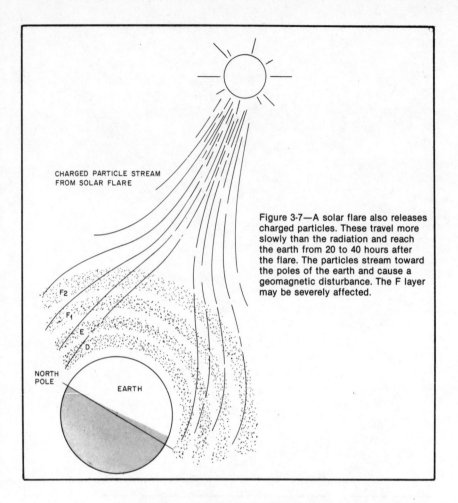

CHARGED PARTICLE STREAM
FROM SOLAR FLARE

Figure 3-7—A solar flare also releases charged particles. These travel more slowly than the radiation and reach the earth from 20 to 40 hours after the flare. The particles stream toward the poles of the earth and cause a geomagnetic disturbance. The F layer may be severely affected.

F2
F1
E
D

NORTH POLE

EARTH

Geomagnetic Disturbances

Another result of a solar flare is a large increase in emission of charged particles from the sun. These particles travel more slowly than the ultraviolet radiation, and reach the earth from 20 to 40 hours after the solar flare. When the charged particles reach the earth's magnetic field they are deflected toward the north and south pole, as shown in Figure 3-7. This means that radio communications along higher-latitude paths (North or South latitudes greater than about 45°) will be more affected than those closer to the equator.

These charged particles affect the upper regions of the ionosphere first—the F layer may seem to disappear or seem to split into many layers. This means that the higher-frequency bands are most affected. The highest frequency that will be returned from the ionosphere may be only half what it would be normally. The charged particles streaming toward the poles also increase auroral activity. In general, a geomagnetic disturbance will degrade long-distance radio communications. Under extreme conditions, the geomagnetic disturbance may completely black out long-distance radio communication, especially on paths that pass near the earth's poles.

[If you are studying for an Element 3B (General) exam, study the questions in subelement 3C in Chapter 11 that begin with numbers 3C-5, 3C-7 and 3C-10.]

The Scatter Modes

When we consider radio propagation, it is convenient to look first at what happens under ideal conditions. All electromagnetic-wave propagation is subject to scattering influences that alter idealized patterns to a great degree, however. The earth's atmosphere, ionospheric layers and any objects in the path of the radio signal act to scatter the energy. If we understand how this scattering takes place, we can use the phenomenon to our advantage.

Forward Scatter

There is an area between the outer limit of ground-wave propagation and the point where the first signals are returned from the ionosphere, as shown in Figure 3-8. We call this area the *skip zone*. This zone is often described as if communications between stations in each other's skip zone were impossible. Actually, because some of the transmitted signal is scattered in the atmosphere, the transmitted signal can be heard over much of the skip zone, if sufficiently sensitive receiving devices and techniques are used. VHF scatter from the troposphere (a region of the atmosphere below the ionosphere) is usable out to about 500 miles from the transmitting station. Ionospheric scatter, mostly from the height of the E region, is most marked at frequencies up to about 60 or 70 MHz, and this type of forward scatter may be discernible in the skip zone out to about 1200 miles. Ionospheric scatter propagation is most noticeable on frequencies above the MUF.

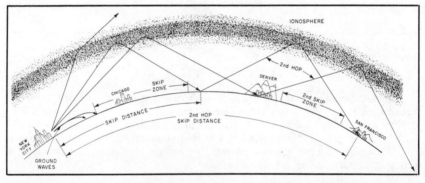

Figure 3-8—There is an area between the farthest reaches of ground-wave propagation and the closest return of sky waves from the ionosphere. This region is known as the skip zone.

A major component of ionospheric scatter is that contributed by meteors as they enter the earth's atmosphere. As the meteor passes through the ionosphere, a column of ionized particles is formed. These particles can act to scatter radio energy. See Figure 3-9. This ionization is a short-lived phenomenon that can show up as short bursts of little communication value or as sustained periods of usable signal level, lasting up to a minute or more. Most common between midnight and dawn, and peaking between 5 and 7 AM local time, meteor scatter can be an interesting adjunct to amateur communication at 21 MHz or higher, especially during periods of low solar activity.

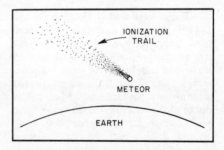

Figure 3-9—Meteors passing through the atmosphere create trails of ionized gas. These ionized trails can be used for short-duration communications.

Backscatter

A complex form of scatter is readily observed when you are working very near the maximum usable frequency. The transmitted wave is refracted back to earth at some distant point, which may be an ocean area or land mass. A small portion of the transmitted signal may be reflected back into the ionosphere toward the transmitter when it reaches the earth. The reflected wave helps fill in the skip zone, as shown in Figure 3-10.

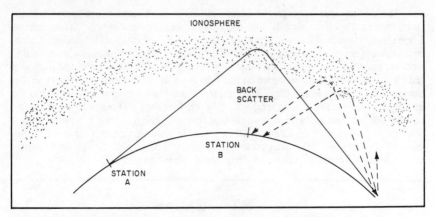

Figure 3-10—When radio waves pass into the ionosphere, some energy may be scattered back into the skip zone.

Backscatter signals are generally rather weak, and subject to some distortion because the signal may arrive at the receiver from many different directions. With optimum equipment, backscatter is usable at distances from just beyond the reliable local range out to several hundred miles. Under ideal conditions, backscatter is possible over 3000 miles or more, though the term "sidescatter" is more descriptive of what probably happens on such long paths.

LINE-OF-SIGHT PROPAGATION

In the VHF and UHF range (above 144 MHz), most propagation is by means of the *space wave*. See Figure 3-11. This space wave is actually made up of a *direct wave,* and usually one or more *reflected waves.* The direct wave and reflected wave arrive at the receiving station slightly out of phase, and may cancel or reinforce each other to varying degrees, depending on the actual path of the reflected wave(s). This effect is most pronounced if the receiving station or the transmitting station (or both) is moving, such as for mobile VHF stations. When one or both stations are moving, the paths are constantly

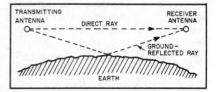

Figure 3-11—Line-of-sight propagation is accomplished by means of the space wave, a combination of a direct ray and one or more reflected rays.

changing, and may alternately reinforce and cancel each other, causing a rapid fluttering sound (called "picket-fencing").

TROPOSPHERIC BENDING AND DUCTING

The troposphere consists of atmospheric layers close to the earth's surface. Although *tropospheric bending* is evident over a wide range of frequencies, it is most useful in the VHF/UHF region, especially at 144 MHz and above. Instead of gradual changes in air temperature, pressure and humidity, distinct layers may form in the troposphere. If two adjacent layers have significantly different densities, radio waves will be bent as they pass from one layer into another, much as light is bent when it passes from air into water or from water into air.

The *true* or *geometric horizon* is the most distant point one can see by line of sight and is limited by the height of the observer above ground. Because of slight bending of the radio waves in the troposphere, however, signals will return to earth somewhat beyond the geometric horizon. This *radio-path horizon* is generally about 15 percent farther away than the true horizon. See Figure 3-12.

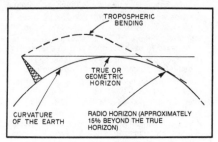

Figure 3-12—Under normal conditions, tropospheric bending causes radio waves to be returned to earth about 15% percent beyond the visual or geometric horizon.

Under normal conditions, the temperature of the air gradually decreases with increasing height above ground. Under certain conditions, however, a mass of warm air may actually overrun cold air so that there is an area above the surface of the earth where cold air is covered by a blanket of warm air. This is called a *temperature inversion*. If the two air masses are more or less stationary, radio waves can be trapped below the warm air, and travel great distances with very little attenuation. The area between the earth and the warm air mass is known as a *duct*. See Figure 3-13.

Radio-wave propagation through a duct is called *guided propagation,* because the manner in which the waves are guided through the duct is very similar to the way microwaves travel through a waveguide. As with tropospheric bending, ducting is primarily observed at VHF frequencies, 144 MHz and higher. Ducts usually form over water, though they can form over land as well. The lowest usable frequency for duct propagation depends on the depth of the duct and the amount the refractive index changes at the air-mass boundary.

[If you are studying for an Element 3A (Technician) exam, turn to Chapter 10 and study questions 3C-6.1, 3C-6.3, 3C-8.1, 3C-8.2 and questions 3C-9.1 through 3C-9.6.

If you are studying for an Element 3B (General) exam, study questions 3C-6.2 and questions 3C-6.4 through 3C-6.6 in Chapter 11.]

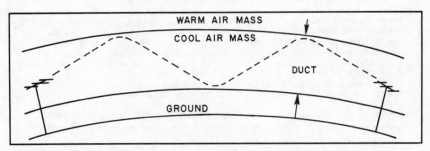

Figure 3-13—When a cool air mass is overrun by a mass of warmer air, a "duct" is formed, allowing VHF radio signals to travel great distances with little attenuation.

Key Words

Antenna noise bridge—a test instrument used to determine the impedance of an antenna system

Audio rectification—interference to electronic devices caused by a strong RF field that is rectified and amplified in the device

Balun—short for BALanced to UNbalanced. The balun is used to transform between an unbalanced feed line and a balanced antenna.

Bleeder resistor—a resistor added across the filter capacitor in a power supply. The bleeder resistor dissipates any charge left on the filter capacitor when the supply is switched off.

Cathode-ray tube (CRT)—a vacuum tube with a phosphor coating on the inside of the face. CRTs are used in oscilloscopes and as the "picture tube" in television receivers.

Dual-trace oscilloscope—an oscilloscope with two separate vertical input circuits. This type of oscilloscope can be used to observe two waveforms at the same time.

Dummy load (dummy antenna)—a resistor that acts as a load for a transmitter, dissipating the output power without radiating a signal. A dummy load is used when testing transmitters.

Feed line—the wires or cable used to connect a transmitter or receiver to an antenna

Field-effect transistor volt-ohm milliammeter (FET VOM)—a multiple-range meter used to measure voltage, current and resistance. The meter circuit uses an FET amplifier to provide a high input impedance for more accurate readings.

Field strength meter—a simple test instrument used to show the presence of RF energy and the relative strength of the RF field

Marker generator—an RF signal generator that produces signals at known frequency intervals. The marker generator can be used to calibrate receiver and transmitter frequency readouts.

Monitor oscilloscope—a test instrument connected to an amateur transmitter and used to observe the shape of the transmitted-signal waveform

Multimeter—an electronic test instrument used to make basic measurements of current and voltage in a circuit. This term is used to describe all meters capable of making different measurements, such as the VOM, VTVM and FET VOM.

Neutralization—feeding part of the output signal from an amplifier back to the input so it arrives out of phase with the input. This negative feedback neutralizes the effect of positive feedback caused by coupling between the input and output circuits in the amplifier.

Oscilloscope—an electronic test instrument used to observe waveforms and voltages on a cathode-ray tube

Radio-frequency interference (RFI)—interference to an electronic device (radio, TV, stereo) caused by RF energy from an amateur transmitter or other source

Reflectometer—a test instrument used to measure the difference between forward power (power from the transmitter) and reflected power (power returned from the antenna system)

RF signal generator—a test instrument that produces a stable low-level radio-frequency signal. The signal can be set to a specific frequency and used to troubleshoot RF equipment.

S meter—a meter in a receiver that shows the relative strength of a received signal

Signal tracer—a test instrument that shows the presence of RF or AF energy in a circuit. The signal tracer is used to trace the flow of a signal through a multistage circuit.

Speech processor—a device used to increase the average power contained in a speech waveform. Proper use of a speech processor can greatly improve the readability of a voice signal.

Sweep circuits—the circuits in an oscilloscope that set the time base for observing waveforms on the scope

Transmatch, or impedance-matching network—a device used to convert the transmitter output impedance (usually 50 ohms) to the impedance of an antenna system

Transmit-receive (TR) switch—a mechanical switch, relay or electronic circuit used to switch an antenna between a receiver and transmitter in an amateur station

Two-tone test—problems in a sideband transmitter can be detected by feeding two audio tones into the microphone input of the transmitter and observing the output on an oscilloscope

Vacuum-tube voltmeter (VTVM)—a multimeter using a vacuum tube in its input circuit. The VTVM does not load a circuit as much as a VOM, and can be used to measure small voltages in low-impedance circuits. A VTVM generally requires 117-V ac power.

Volt-ohm-milliammeter (VOM)—a test instrument used to measure voltage, resistance and current. A VOM usually has a RANGE switch so the meter can be used to measure a wide range of inputs.

Wattmeter—a test instrument used to measure the power output (in watts) of a transmitter

Chapter 4

Amateur Radio Practice

I t's easy to form good operating habits in Amateur Radio. By understanding the fundamentals of radio, you should be able to maintain your station so that its operation is in accordance with FCC rules and regulations. In addition, if you're like most amateurs, you'll want to put out the best possible signal. If you know the principles behind the operation of your station, correctly adjust and measure its performance and take advantage of available test equipment, you should have no problem keeping your station "up to par."

AMATEUR RADIO SAFETY

Your station equipment not only makes use of ac line voltage, which can be dangerous in itself—it also generates additional potentially lethal voltages of its own. For your own safety, and that of others who may come in contact with your equipment, you should be familiar with some basic precautions.

Power-Line Connections

In most residential systems, three wires are brought in from the outside to the distribution board. Older systems may use only two wires. In the three-wire system, the voltage between the two "hot" wires is normally 234. The third wire is neutral and is grounded. Half of the total voltage appears between each of the hot wires and neutral, as indicated in Figure 4-1. In systems of this type, the 117-V outlets,

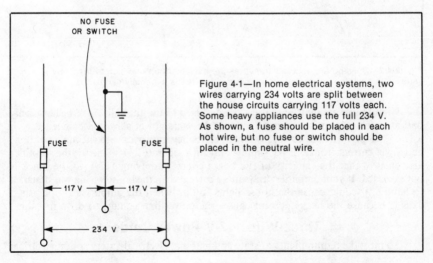

Figure 4-1—In home electrical systems, two wires carrying 234 volts are split between the house circuits carrying 117 volts each. Some heavy appliances use the full 234 V. As shown, a fuse should be placed in each hot wire, but no fuse or switch should be placed in the neutral wire.

lights and appliances are divided as evenly as possible between the two sides of the circuit. Half of the load is connected between one hot wire and the neutral, and the other half of the load is connected between the other hot wire and neutral.

Heavy appliances, such as electric stoves and most high-power amateur amplifiers, are designed for 234-V operation and are connected across the two hot wires. While **both** ungrounded wires should be fused, a fuse or switch should **never** be used in the neutral wire. If the appliance is connected to the supply voltage with a four-conductor power cord, the black and red wires should be connected to the fused hot wires. The white and green (or bare) wires should not have a fuse or switch in the line. The reason for this is that opening the neutral wire does not disconnect the equipment from the household voltage. It simply leaves the equipment on one side of the 234-V line in series with whatever load may be across the other side of the circuit, as shown in Figure 4-2. Furthermore, with the neutral wire open, the voltage will then be divided between the two sides in inverse proportion to the load resistance. The voltage will drop below normal on one side and soar on the other, unless both loads happen to be equal.

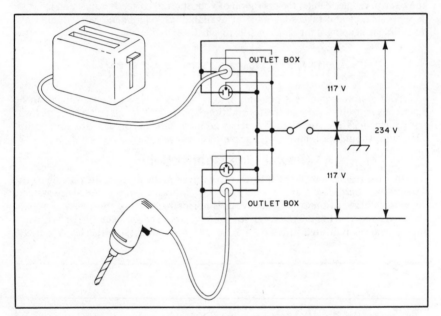

Figure 4-2—Opening the neutral wire leaves the equipment on one side of the 234-V line in series with anything connected on the other side. This is extremely dangerous!

High-power amateur amplifiers are designed to use the full 234 V because only half as much current is required to supply the same power as would be needed with a 117-V supply. The power-supply circuit efficiency improves with the higher voltage and lower current from a 234-V supply. Having a separate 234-V line for the amplifier also ensures that the capacity of the house circuit supplying 117 V to the shack is not exceeded. If your amplifier draws more current than the house wiring was designed to handle, it could cause the house lights to dim when the amplifier is operating. This is because the heavy current causes the power-line voltage to drop in value.

Three-Wire 117-V Power Cords

To meet the requirements of state and national electrical-safety codes, electrical tools, appliances and many items of electronic equipment now being manufactured to operate from the 117-V line must be equipped with a three-conductor power cord. Two of the conductors (the "hot" and "neutral" wires) carry power to the device, while the third conductor (the ground wire) is connected to the metal frame of the device. See Figure 4-3. The "hot" wire is usually black or red, the "neutral" wire is white, and the frame/ground wire is green, or sometimes bare.

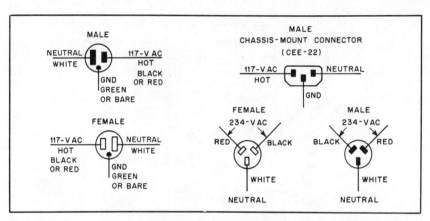

Figure 4-3—Correct wiring technique for 117-V and 234-V power cords and receptacles. The white wire is neutral, the green wire is ground and the black or red wire is the hot lead.

When plugged into a properly wired mating receptacle, a three-contact plug connects the third conductor to an earth ground, thereby grounding the chassis or frame of the appliance and preventing the possibility of electric shock to the user. A defective power cord that shorts to the case of the appliance will simply blow a fuse. Without the ground connection, the case could carry the full line voltage, presenting a severe shock hazard. All commercially manufactured electronic test equipment and most ac-operated amateur equipment is being supplied with these three-wire cords. Adapters are available for use where older electrical installations do not have mating receptacles. For proper grounding, the lug of the green wire protruding from the adapter must be attached underneath the screw securing the cover plate of the outlet box. The outlet itself must be grounded for this to be effective. If the power wires coming into the electric box are inside a flexible metal covering, the outlet should be grounded through the metal covering. This type of wire is commonly referred to as armored cable.

A "polarized" two-wire plug and mating receptacle ensures that the hot wire and the neutral wire in the appliance are connected to the appropriate wires in the house electrical system. A polarized plug has one blade that is wider than the other. Without this polarized plug and receptacle, the power switch in the equipment may be in the hot wire when the plug is inserted one way, and in the neutral wire when inserted the other way. This can present a dangerous situation. It is possible for the equipment to be "hot" even with the switch off. With the switch in the neutral line, the hot line may be connected to the equipment chassis. An unsuspecting operator could form a path to ground by touching the case, and might receive a nasty shock!

When wiring an outlet or lamp socket, it is important to connect the wires properly. As shown in Figure 4-4, the black (hot) wire should be connected to the brass terminal on the lamp socket or outlet, and the white (neutral) wire should be connected to the white-colored terminal on the socket or outlet. This will ensure that

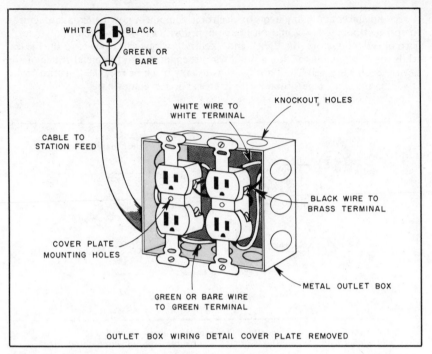

WHITE | BLACK
GREEN OR BARE
WHITE WIRE TO WHITE TERMINAL
KNOCKOUT HOLES
CABLE TO STATION FEED
BLACK WIRE TO BRASS TERMINAL
COVER PLATE MOUNTING HOLES
METAL OUTLET BOX
GREEN OR BARE WIRE TO GREEN TERMINAL

OUTLET BOX WIRING DETAIL COVER PLATE REMOVED

Figure 4-4—The correct way to wire a receptacle box. Connect the white wire to the white terminal, and the black wire to the brass-colored terminal. This ensures that the prongs on the mating plug will be connected properly.

the proper blade of the plug will connect to the hot wire. This practice is especially important when wiring lamp sockets. The brass screw of the socket is connected to the center pin in the socket. With this pin as the "hot" connection, the shock hazard is greatly reduced. A person unscrewing a bulb from a correctly wired socket would have to reach down inside the socket to get a shock—in an incorrectly wired socket, the screw threads of the bulb are "hot" and a dangerous shock may result when replacing a bulb.

Current Capacity

Another factor that must be taken into account when you are wiring an electrical circuit is the current-handling capability of the wire. Table 4-1 shows the current-handling capability of some common wire sizes. From the table we can see that number 14 wire could be used for a circuit carrying 15 A, while number 12 (or larger) would have to be used for a circuit carrying 20 A.

To remain safe, don't overload the ac circuits in your home. The circuit breaker or fuse rating indicates the

Table 4-1
Current-Carrying Capability of Some Common Wire Sizes

Wire Size AWG	Continuous-Duty Current*
8	46 A
10	33 A
12	23 A
14	17 A
16	13 A
18	10 A
20	7.5 A
22	5 A

*wires or cables in conduits or bundles

maximum load that may be placed on the line at any one time. Simple mathematics can be used to calculate the current your ham radio equipment will draw. Most equipment has the power requirements printed on the back. If not, the owner's manual should contain such information. When calculating current requirements, make sure to include any other household appliances that may be on the same line—including lights! If you were to put a larger fuse in the circuit, too much current could be drawn, the wires would become hot and a fire could result.

[If you are studying for an Element 3A (Technician) exam, turn to Chapter 10 and study questions 3D-1.1 through 3D-1.4.

If you are studying for an Element 3B (General) exam, study questions 3D-1.5 through 3D-1.9 in Chapter 11.]

Power-Supply Safety

Safety must always be carefully considered during the design and construction of any power supply. Power supplies can produce potentially lethal voltages, and care must be taken to guard against accidental exposure. For example, electrical tape, insulated tubing (spaghetti) or heat-shrink tubing is recommended for covering exposed wires, component leads, component solder terminals and tie-down points. Whenever possible, connectors used to mate the power supply to the outside world should be of an insulated type designed to prevent accidental contact with the voltages present. Ac power to the supply should be controlled by a clearly labeled switch, located on the front panel where it can be seen and reached easily in an emergency.

All dangerous voltages in equipment should be made inaccessible. A good way to ensure this is to enclose all equipment in metal cabinets so that no "hot" spots can be reached. Don't forget any component shafts that might protrude through the front panel; if hot, these may be protected from contact by means of an insulated shaft extension or insulated knob.

Each metal enclosure should be connected to a good earth ground, such as a ground rod. Thus, if a failure occurs inside a piece of equipment, the metal case will never present a shock hazard; the fuse will blow instead.

It's also a good idea to make it impossible for anyone to energize your equipment when you're not present. A key-operated ac-mains switch to control all of the power to your station is a good way to accomplish this. Whatever type of switch you use, be sure you mount it where it can be seen and reached easily in an emergency.

You should never underestimate the potential hazard when working with electricity. Table 4-2 shows some of the effects of electric current—as little as 100 mA can be lethal! As the saying goes, "It's volts that jolts, but it's mils that kills." Low-voltage power supplies may seem safe, but even battery-powered equipment should be treated with respect. Thirty volts is the minimum voltage considered dangerous to humans, but these voltage and current ratings are only general guidelines. Automobile batteries are designed to provide very high current (as much as 200 A) for short periods when starting a car, and this much current can kill you, even at 12 volts. You will feel pain if the shock current is in the range of 30 to 50 mA.

Table 4-2

Effects of Electric Current Through the Body of an Average Person

Current (1 Second Contact)	Effect
1 mA	Just perceptible.
5 mA	Maximum harmless current.
10-20 mA	Lower limit for sustained muscular contractions.
30-50 mA	Pain
50 mA	Pain, possible fainting. "Can't let go" current.
100-300 mA	Normal heart rhythm disrupted. Electrocution if sustained current.
6 A	Sustained heart contraction. Burns if current density is high.

A few factors affect just how little voltage and current can be considered dangerous. One factor is skin resistance—the lower the resistance of the path, the more current that will pass through it. If you perspire heavily, you may get quite a bit more severe shock than if your skin was dry. Another factor is the path through the body to ground. As shown in Figure 4-5, the most dangerous path is from one hand to

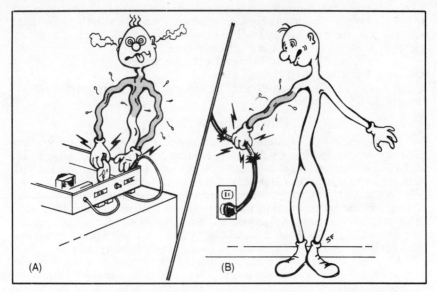

Figure 4-5—The path from the electrical source to ground affects how severe an electrical shock will be. The most dangerous path (from hand to hand directly through the heart) is shown at A. The path from one finger to the other shown at B is not quite so dangerous.

the other. This path passes directly through a person's heart, and even a very minimal current can cause heart failure and death. By contrast, current passing from one finger to another on the same hand will not have quite such a serious effect. For this reason, if you must troubleshoot a live electrical circuit, it's best to keep one hand behind your back or in your pocket. If you do slip, the shock may not be as severe as if you were using both hands.

Bleeder Resistors

An important safety item in power-supply design is the *bleeder resistor*. When a power supply is turned off, the filter capacitors can store a charge for a long time, presenting a shock hazard at the output terminals. Connected across the filter capacitor, a bleeder resistor will dissipate the charge stored in the capacitor when the supply is turned off. This will not affect normal operation of the supply because the bleeder resistor draws only a very small current.

[If you are studying for an Element 3A (Technician) exam, turn to Chapter 10 and study questions 3D-2.1, 3D-2.3 and 3D-2.5.

If you are studying for an Element 3B (General) exam, study question 3D-2.4 in Chapter 11.]

USING TEST EQUIPMENT

The FCC requires applicants for the General class license to be familiar with the use of four pieces of test equipment: oscilloscope, multimeter, RF signal generator and signal tracer. Many of you may have already used most of these items.

The Oscilloscope

The *oscilloscope* (sometimes called a scope), is probably the most versatile of all instruments. Oscilloscopes measure instantaneous voltage changes by the vertical deflection of an electron beam as it is swept horizontally across the face of a *cathode-ray tube (CRT)*. An oscilloscope has amplifiers in line with its horizontal and vertical inputs, and can respond to a wide range of frequencies. Expensive oscilloscopes can handle frequencies over 30 MHz, while the scopes most amateurs use only respond well up to 5 MHz or so. This lets you display audio- or low radio-frequency signals.

Oscilloscopes contain *sweep circuits,* which are oscillators used to draw the trace horizontally across the screen. The sweep-oscillator frequency must be adjusted to the proper range to display various signal frequencies. For Amateur Radio use, a properly adjusted scope can display the output waveform of your transmitted signal, as shown in Figure 4-6. This can be especially useful if you wish to check the modulation level of your signal. It can also show if there is carrier present on your single-sideband (SSB) signal. Another use for the oscilloscope is to display circuit frequency response. Morse code elements can be displayed as shown in Figure 4-7, giving a graphic demonstration of the rise and fall times of the transmitted waveform.

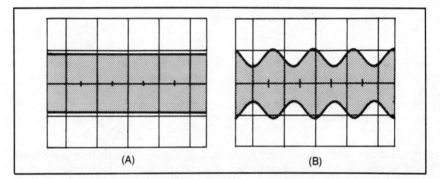

Figure 4-6—Oscilloscope displays of RF signals. At A is an unmodulated carrier. The signal at B is from a full-carrier AM transmitter modulated with a single-frequency sine wave.

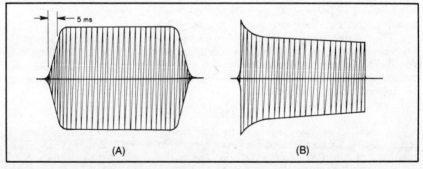

Figure 4-7—At A is the waveform of a single CW dot as seen on a monitor oscilloscope. Note the rounded corners and the smooth rise and decay times. Five milliseconds is generally agreed to be the ideal rise and fall time of a CW envelope. The waveform at B is typical of a poorly designed and adjusted transmitter. The sharp rise and decay transitions contain energy at frequencies other than the desired frequency.

Two signals can be compared to each other using an oscilloscope. Apply one signal to the vertical input and the other to the horizontal input. Depending on the frequency relationship between the two inputs, a particular pattern will be displayed

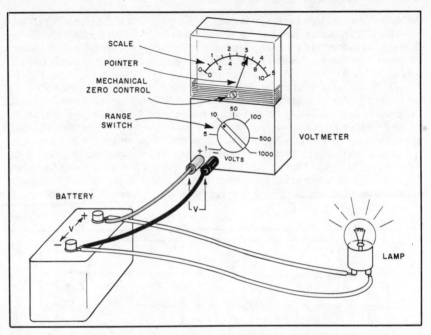

SCALE
POINTER
MECHANICAL
ZERO CONTROL
RANGE
SWITCH
VOLTMETER
BATTERY
LAMP

Figure 4-8—When you use a multimeter to measure voltage, the meter must be connected in parallel with the voltage you want to measure.

on the screen. This pattern is called a Lissajous figure. You will learn more about how to interpret Lissajous figures as you study for your Advanced class license.

Another useful comparison can be made using a *dual-trace oscilloscope*. This type of scope has two separate vertical input channels. The two channels can be swept in the horizontal direction at the same rate, and if two different signals are applied to the two separate vertical inputs, both signals can be viewed on the screen at the same time. This function is especially useful for comparing the characteristics of a signal before and after passing through a stage in a circuit.

There are a variety of controls on any oscilloscope, and you should follow the instructions in the manual that comes with the scope when adjusting them. The focus control should be carefully adjusted for a sharp, fine line on the CRT. The intensity control sets the brightness of the trace. Setting the brightness too high can damage the face of the oscilloscope CRT.

The Monitor Scope

A *monitor oscilloscope* is used to show the signal quality of a transmitter. It lets you know if your SSB signal is flattopping, what your CW dot-dash ratio is and the shape of your CW signal waveform. The vertical input of the monitor scope is connected to the transmitter output. A clean CW signal is shown in Figure 4-7A. Notice the smooth rise and decay. A CW signal with key clicks is shown in Figure 4-7B.

The Multimeter

A *multimeter* is a piece of test equipment that most amateurs should know how to use. The simplest kind of multimeter is the *volt-ohm-milliammeter (VOM)*. VOMs use one basic meter movement for all functions; the movement requires a fixed amount of current (often 1 mA) for a full-scale reading. A switch selects various ranges for voltage, resistance and current measurements. This switch places high-value dropping resistors in series with the meter movement for voltage measurement, and connects low-value shunt resistors in parallel with the movement for current measurement. These parallel and series resistors extend the range of the basic meter movement.

When you use a multimeter for voltage measurement, connect the meter terminals in parallel with the voltage to be measured, as shown in Figure 4-8. An ideal voltmeter would have an infinite input impedance and would not affect the circuit it is connected to in any way. Real-world voltmeters, however, have a finite value of input impedance. There is a possibility that the voltage to be measured will change because of additional circuit loading when the voltmeter is connected.

It is a good idea to use a voltmeter with an input impedance that is very high compared to the impedance of the circuit being measured. This prevents the voltmeter from drawing excessive current from the circuit. Excessive current drawn from the circuit would significantly affect its operation. The input impedance of an ordinary voltmeter is on the order of 20 kilohms per volt. This means that the internal resistance of the meter is equal to the range setting in volts multiplied by 20,000.

If you are going to purchase a VOM, buy one with the highest ohms-per-volt rating that you can find. Stay away from meters rated much under 20,000 ohms per volt if you can.

When you use a multimeter for current measurement, connect the meter terminals in series with the current to be measured, as shown in Figure 4-9. Most ammeters have very low resistance, but in a low-impedance circuit there is a chance that the slight additional resistance of the series ammeter will disturb circuit operation.

Measuring resistance with a meter involves placing the meter leads across the component or circuit you wish to measure. Make sure to select the proper resistance scale. The full-scale-reading multipliers vary from 1 to 1000 and higher, and the scale is usually compressed on the higher end of the range, as shown in Figure 4-10. For

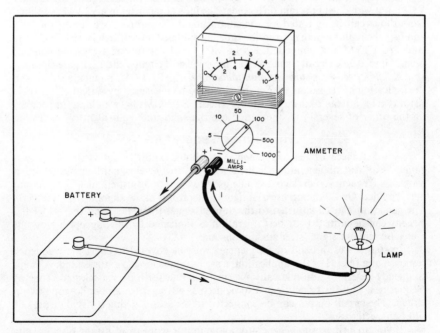

Figure 4-9—To measure current you must break the circuit at some point and connect the meter in series at the break.

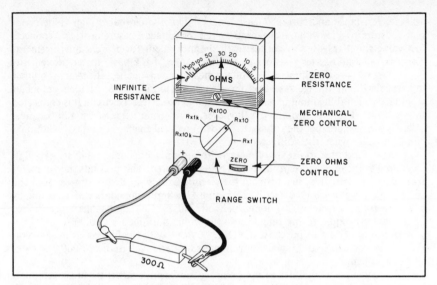

Figure 4-10—Resistance is measured across the component. For best accuracy the reading should be taken in the lower half of the scale.

best accuracy, you need to keep the reading in the lower half of the scale (on most meters, the right-hand side). Thus, if you want to measure a resistance on the order of 5000 ohms, you would select the R × 1000 scale.

A *vacuum-tube voltmeter (VTVM)* operates in the same manner as an ordinary VOM, but with one important difference: The indicating meter in a VTVM is isolated from the circuit being tested by a vacuum-tube dc amplifier. As a result, the only additional circuit loading is from the tube input impedance, which is very high. The standard VTVM input impedance is 11 MΩ. This is useful for measuring voltages in high-impedance circuits, such as vacuum-tube grid circuits and FET gate circuits.

A *field-effect transistor volt-ohm-milliammeter (FET VOM)* is manufactured with the indicating meter isolated from the circuit to be measured by a field-effect transistor (FET), instead of a vacuum tube. These FET VOMs have an input impedance on the order of several megohms, and are the solid-state equivalent of a VTVM.

RF Signal Generators

Another piece of test equipment you should know about is the *RF signal generator*. A signal generator produces a stable, low-level signal that can be set to a specific frequency. This signal can be used to align or adjust circuits for optimum performance. One common use of a signal generator is in the alignment of receivers. The generator can be adjusted to the desired signal frequency, and the associated circuits can be adjusted for best operation as indicated by the appropriate output meter (maximum signal strength, for instance).

In some cases, a band of frequencies must be covered to chart the frequency response of a filter. For this application, a swept frequency generator is used. A swept frequency generator automatically sweeps back and forth over a selected range of frequencies. This instrument can also be used as a trigger for an oscilloscope sweep circuit. The circuit output can be connected to the scope, giving a visual display of filter characteristics.

If you are using a signal generator and monitor scope to adjust the filter circuits in your transmitter, a *dummy load* must be connected to the transmitter output. The dummy load acts as a constant load for the transmitter, replacing the antenna without radiating a signal. More about dummy loads later.

The *marker generator* is a special type of RF signal generator. In its simplest form, the marker generator is a high-stability oscillator generating a series of signals that, when detected in the receiver, mark the exact edges of the amateur bands (and subbands, in some cases). The marker generator does this by oscillating at a low frequency that has harmonics falling on the desired frequencies. Most marker generators put out harmonics at 25, 50 or 100-kHz intervals. Since marker generators normally use crystal oscillators they are often called crystal calibrators.

The marker generator is very useful for determining transmitter frequency. The signal from the transmitter is first tuned in on the receiver, and the dial setting is noted. Then, with the marker generator turned on, the closest markers above and below the transmitter frequency are noted. The transmitter frequency must be between the two markers.

If the marker frequencies are accurate, all that needs to be known is that the transmitter frequency is above the marker that shows the low end of the band (or subband) and below the marker for the high end of the band (or subband). In addition, the transmitter frequency must not be so close to the edge of the band (or subband) that the sideband frequencies, especially in a voice transmission, will extend over the edge.

The Signal Tracer

With today's multistage rigs, most problems that come about are a result of a defective component in only one stage. Isolating the problem stage is the first step in the repair process. Armed with a block diagram of your radio and a *signal tracer,* you can proceed to locate the trouble spot in a straightforward manner.

A signal tracer, as the name implies, is a device that allows you to trace the path of a signal through the various stages in your rig until it disappears or begins to show unusual characteristics. Any device that reacts to RF or AF energy can be used as a signal tracer. A typical signal tracer consists of a diode detector and a high-gain audio amplifier, which detects amplitude variations in RF signals. An oscilloscope or a VTVM can also be used to isolate trouble in your ham equipment.

[If you are studying for an Element 3A (Technician) exam, turn to Chapter 10 and study questions 3D-7.1 through 3D-7.4 and questions 3D-8.1 through 3D-8.5.

If you are studying for an Element 3B (General) exam, study questions 3D-6.1 through 3D-6.4, 3D-9.1 through 3D-9.3, 3D-15.3 and 3D-15.4 in Chapter 11.]

TESTING TRANSMITTER PERFORMANCE

Some of the most important tests and measurements you make as an amateur are those that tell you about the quality of your transmitted signal. Other hams form their first opinions of you on the air from the quality of your signal. Improperly adjusted amateur transmitters wreak havoc all over the bands and make operating unpleasant for everyone. An improperly adjusted transmitter can radiate energy all over the spectrum—not just in the amateur bands! Unless you know how to monitor your transmitted signal you are asking for problems with radio-frequency interference (RFI) and television interference (TVI).

Two-Tone Tests

The *two-tone test* is used to check the amplitude linearity of an SSB transmitter. Single-sideband suppressed-carrier signals are given the emission designator J3E. Instead of the continuously varying pattern produced by a voice signal, a two-tone test produces a stationary pattern that can be examined on an oscilloscope. Figure 4-11A shows a pattern that is representative of a properly operating SSB

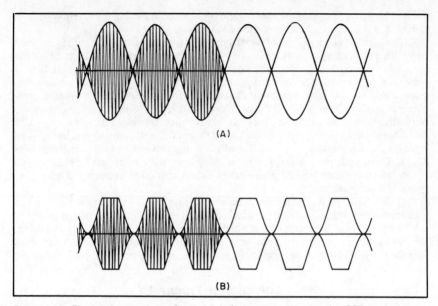

(A)

(B)

Figure 4-11—The two-tone test waveform at A is from a properly operating SSB transmitter; note the smooth sinusoidal shape of the curves, especially as they cross the zero axis. At B is the waveform of a two-tone test showing severe flattopping and crossover (zero-axis) distortion.

transmitter. Figure 4-11B represents a signal that exhibits flattopping and crossover distortion. In Figure 4-12, a single cycle of a sine-wave signal is compared with a cycle of a signal that exhibits flattopping and crossover distortion.

To perform a two-tone test, you inject two audio signals of equal level into the transmitter microphone jack. The signals used should be about 1 kHz apart for a good display on the monitor oscilloscope and must be within the audio passband of the transmitter. Do not use frequencies that are harmonically related (like 1000 and 2000 Hz). Tones at 700 Hz and 1900 Hz are used in the ARRL lab for equipment testing.

In a "clean" transmitter, this two-tone input will produce an RF signal at the output of the transmitter that contains only the two input signals. No amplifier is perfectly

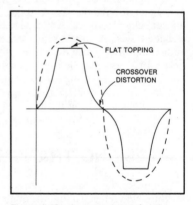

Figure 4-12—A comparison of a sine-wave signal with a signal that exhibits flattopping and crossover distortion.

linear, so some mixing of the two input signals will take place. The sum and difference products produced by this mixing should be very weak in comparison with the main output. They should be weak enough that they cannot be detected on an oscilloscope pattern. What you will see on an oscilloscope is the pattern of two sine-wave signals as they add and subtract, forming peaks and valleys, as shown in Figure 4-11A.

A spectrum analyzer will tell you much more about your transmitter output, but such instruments are expensive, and few amateurs have access to one for transmitter testing. Oscilloscopes are relatively inexpensive and readily available. A two-tone test pattern observed on a scope will show only major defects in the output of an SSB transmitter, but it is still a very useful indication of how the transmitter is operating. If you keep a monitor oscilloscope connected to your transmitter output at all times you will be able to quickly notice any changes in operation. The speech pattern from a correctly adjusted SSB transmitter (emission J3E) is shown in Figure 4-13A. Notice that the peaks are smoothly rounded. Overdriving the transmitter will cause the waveform to look more like the one shown in Figure 4-13B, with the peaks flattened, or clipped.

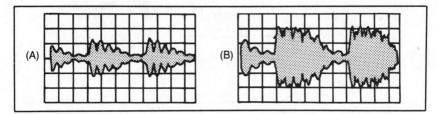

Figure 4-13—The speech pattern from a correctly adjusted sideband transmitter is shown at A. B shows the same transmitter with excessive drive, causing peak clipping in the final amplifier.

Neutralization of Power Amplifiers

Most RF amplifiers operate with their input and output circuits tuned to the same frequency. Unless the circuit is carefully designed, some of the amplified output signal may be fed back to the input. If the signal is fed back "in phase" (so it adds to the input signal), the amplifier will oscillate. Care should be used in arranging components and wiring of the input and output circuits so that coupling is kept to a minimum. Keep all RF leads as short as possible, and pay particular attention to the RF return paths from input and output tank circuits to the emitter, drain or cathode.

Even with proper shielding and careful design, interelectrode capacitance in vacuum-tube amplifiers can cause problems. In modern vacuum tubes the internal elements are so close together that signals can couple from the plate back to the grid, causing the amplifier to oscillate. This interelectrode capacitance is represented by C_{gp} in Figure 4-14. With tetrode and pentode tubes, interelectrode capacitance is greatly reduced by the addition of a screen grid between the control grid and the plate. Nevertheless, the power sensitivity of these tubes is so great that only a small amount of feedback is necessary to start oscillation.

In many cases, a technique called *neutralization* is used to provide for stable amplifier operation. Neutralization is the process of deliberately feeding a portion of the amplifier output back to the input 180° out of phase with the input. (This is called negative feedback because the feedback signal subtracts from the input.) The feedback is usually accomplished by the addition of a neutralizing capacitor connected to the tank circuit (C_n in Figure 4-14). If done correctly, neutralization will cancel any positive feedback in the plate-grid capacitance. This will render the amplifier much more stable and make it less likely to oscillate.

Neutralizing a Screen-Grid Amplifier Stage

One way to accomplish neutralization is to reduce to a minimum the RF-driver voltage fed from the input of the amplifier to its output circuit through the grid-

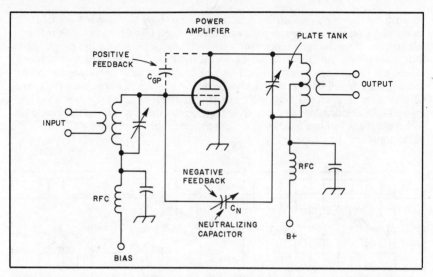

Figure 4-14—In most vacuum-tube amplifiers, some form of neutralization must be used to cancel positive feedback from the plate to the grid. In this circuit, a neutralizing capacitor (C_n) is added to counteract the effects of the grid-to-plate capacitance (represented by C_{gp}). The signal at the bottom of the tank circuit is 180° out of phase with the signal at the top of the tank, and careful adjustment of C_n will effectively cancel the positive feedback.

plate capacitance of the tube. If the screen-grid tube is operated without grid current, a sensitive output indicator can be used to indicate neutralization. With this technique, both screen and plate voltages must be removed from the tube(s). The dc circuits from the plate and screen to cathode are left intact. The neutralizing capacitor or link coils are then carefully adjusted until the output indicator reads minimum.

If the tube is operated with grid current, the grid-current meter can be used to indicate neutralization. With this technique the plate voltage may remain on, but the screen voltage must be zero, with the dc circuit completed between screen and cathode. There will be a change in grid current as the unloaded plate tank circuit is tuned through resonance. The neutralizing capacitor (or inductor) should be adjusted until this deflection is brought to a minimum.

As a final adjustment, screen voltage should be reapplied and the neutralizing adjustment continued to the point where minimum plate current, maximum grid current and maximum screen current occur simultaneously. An increase in grid current when the plate tank circuit is tuned slightly to the high-frequency side of resonance indicates that the neutralizing capacitance is too small. If the increase is on the low-frequency side, the neutralizing capacitance is too large. When neutralization is complete, there should be a slight decrease in grid current on either side of resonance.

[If you are studying for an Element 3B (General) exam, turn to Chapter 11 and study questions 3D-3.1 through 3D-3.5, questions 3D-4.1 through 3D-4.5, 3D-15.1 and 3D-15.2.]

STATION ACCESSORIES

There are a number of station accessories that make the difference between simply operating a radio and having a full awareness of the communications medium called Amateur Radio. In this section we will look at some of the accessories you may find useful when you upgrade to a Technician or General class license.

The S Meter

Because reception is (or should be) the first aspect of station operation, we'll examine the *S meter* first. S meters are used on receivers to indicate the strength of a received signal. Signal-strength meters are useful when there is a need to make comparative readings.

S meters are calibrated in S units from S1 to S9; above S9 they are calibrated in decibels (dB). (You will learn more about decibels in Chapter 5.) An attempt was made by at least one manufacturer in the 1940s to establish some significant numbers for S meters. S9 was to be equal to a signal level of 50 microvolts, with each S unit equal to 6 dB. This means that for an increase of one S unit, the received signal power would have to increase by a factor of four. Such a scale is useful for a theoretical discussion, although real S-meter circuits fall far short of this ideal for a variety of reasons.

Suppose you were working someone who was using a 25-watt transmitter, and your S meter was reading S7. If the other operator increased power to 100 watts (an increase of four times), your S meter would read S8. If the signal was S9, the operator would have to multiply his power 10-fold to make your S meter read 10 dB over S9.

S meters on modern receivers may or may not respond in this manner. S-meter operation is based on the output of the automatic gain-control (AGC) circuitry; the S meter measures the AGC voltage. As a result, every receiver S meter responds differently; no two S meters will give the same reading. The S meter is useful for giving relative signal-strength indications, however. You can see changes in signal levels on an S meter that you may not be able to detect just by listening to the audio output level.

The Field-Strength Meter

A *field-strength meter* is a simple instrument that measures the relative strength of an RF field. This makes it useful for monitoring the radiated signal strength during antenna and transmitter adjustments. A field-strength meter usually contains a diode detector and a milliammeter (to indicate signal strength). An antenna is attached to the detector input. By placing a field-strength meter a fixed distance from a rotatable antenna, you can get a relative indication of the antenna radiation pattern by observing variations in signal strength as the antenna is rotated.

The Dummy Load

A *dummy load* (or dummy antenna, as it is sometimes called) is a resistor that has impedance characteristics that allow it to take the place of a regular antenna system for test purposes. With a dummy load connected to the transmitter, you can make tests without having a signal go out over the air. A dummy load is especially handy for transmitter testing, as FCC rules strictly limit the amount of on-the-air testing that may be done (see Section 97.93 of the amateur rules).

For transmitter tests, remember that a dummy load is a resistor and must be capable of safely dissipating the entire power output of the transmitter. A dummy load should provide the transmitter with a perfect load. This usually means that it has a pure resistance (with no reactance) of approximately 50 ohms. The resistors used to construct dummy loads must be noninductive. Composition resistors are usable, but wire-wound resistors are not. If a single high-power resistor is not available, several lower-power resistors can be connected in parallel to obtain a 50-ohm load capable of dissipating high power.

[If you are studying for an Element 3A (Technician) exam, turn to Chapter 10 and study questions 3D-16.1 through 3D-16.5 and 3D-17.1.

If you are studying for an Element 3B (General) exam, study questions 3D-17.2 through 3D-17.6 in Chapter 11.]

TR Switching

The *transmit-receive (TR) switch* connects the antenna to the transmitter during periods of sending, and to the receiver during periods of reception. The TR switch may be a manually operated switch, an antenna relay that is switched by the transmitter or receiver, or an electronic circuit. When an electronic TR switch is used, the transmitter is continuously connected to the antenna. The receiver antenna connection is made through the TR switch. During transmit periods, the TR switch prevents large amounts of RF from reaching the receiver by electronically disconnecting the receiver from the antenna.

Since no mechanical device is involved in the antenna switching, the TR switch can follow CW keying. The receiver is left operative continuously, and if it can recover from a strong signal quickly enough, it will be sensitive to received signals between the transmitted character elements. This can provide what is called "full-break-in (QSK) operation" where the operator is able to hear signals in between the dots and dashes of his code transmission.

The tube or transistor that performs the switching function in a TR switch may be driven into nonlinearity by the strong transmitter signal. This is why some electronic TR switches can generate harmonics that may cause TVI. A low-pass filter placed between the TR switch and the antenna, as shown in Figure 4-15, may eliminate the TVI, but the low-order harmonics (say, the second harmonic of 40 meters) may cause interference to other communications (on 20-meter SSB, for instance).

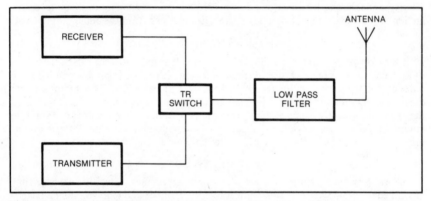

Figure 4-15—A low-pass filter connected between the TR switch and the antenna will help filter out harmonics generated in the TR switch.

Speech Processors

The human voice does not have a constant amplitude. When you speak, the speech amplitude is ever-changing, and only occasionally reaches a maximum, or peak, value. As can be seen in Figure 4-16A, there are large "valleys" between the voice peaks—periods where there is very little voice energy, hence little transmitter output. Because of the valleys in the speech waveform, the average power contained in a voice signal is small, while the peak power is large. Peak-to-average ratios of 2 or 3 are quite common. Using a *speech processor,* the amplitude of the valleys can be increased while leaving the voice peaks unchanged. See Figure 4-16B. This serves to increase the average power contained in the waveform and improve the signal-to-noise ratio, but can lead to excessive background noise pickup. Properly used, speech processing can improve speech intelligibility at the receiver when propagation conditions are poor or the interference level is high.

The increased average power contained in a processed signal places a higher power

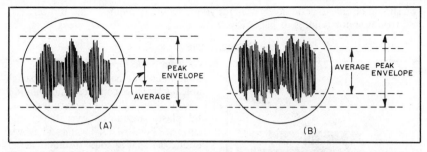

Figure 4-16—A typical SSB voice-modulated signal might have an envelope of the general nature shown at A, where the RF amplitude (current or voltage) is plotted as a function of time, which increases to the right. The envelope pattern after speech processing to increase the average level of power output is shown at B.

demand on the transmitter. This can result in overheating of final amplifier and power supply components. If overheating becomes evident, you may need to reduce the processing level or transmitter power.

Speech processing, by its very nature, is a distortion of the audio signal, so processed signals sound less natural at the receiving end than unprocessed ones. For these reasons, processing should be employed only when necessary. If the frequency is clear and propagation is good, keep your speech processor turned off. If you do need the processor, use the minimum amount of processing necessary to increase intelligibility for the receiving station; overprocessed, distorted signals are more difficult to copy than unprocessed ones. Distortion products generated during speech processing can increase the bandwidth of your signal and cause "splatter" (interference to stations on nearby frequencies). Keep in mind that speech processing cannot increase the peak power of the signal, even if the transmitter is 100% modulated.

Two methods of speech processing can be employed. They are classified as to where in the transmitter the processing occurs.

An AF speech processor acts on the voice signal while it is still at audio frequency. This type of processor is usually an external accessory, and is inserted between the microphone and the transmitter. Such a processor often has more gain than is necessary, so you must be careful not to overdrive the microphone input of your transmitter.

An RF speech processor works on the voice signal after it has been converted to a radio-frequency signal by the transmitter. This processor is often built into a transmitter when it is manufactured. Again, care in adjustment and operation is necessary; you should follow the manufacturer's operating instructions.

[If you are studying for an Element 3A (Technician) exam, turn to Chapter 10 and study question 3D-12.1.

If you are studying for an Element 3B (General) exam, turn to Chapter 11 and study questions 3D-12.2 through 3D-12.5, question 3D-13.1 and question 3D-13.2.]

ANTENNA MEASUREMENTS

A properly operating antenna system is essential for a top-quality amateur station. If you build your own antennas, you must be able to tune them for maximum operating efficiency. Even if you buy commercial antennas, there are tuning adjustments that must be made because of differences in mounting location. Height above ground and proximity to buildings and trees will have some effect on antenna operating characteristics.

You must be able to measure your antenna's performance. After installation, you should be able to periodically monitor your antenna system for signs of problems, and troubleshoot failures as necessary.

Feed Lines

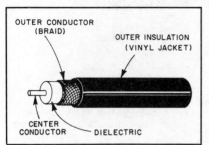

Figure 4-17—Coaxial cable construction is shown.

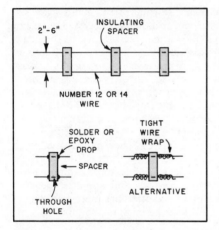

Figure 4-18—"Ladder line" is constructed of two parallel wires separated at intervals by insulators.

Two basic types of *feed line* are in common use at amateur stations. Coaxial cable has a center conductor surrounded by insulation, a shield braid around that, and plastic insulation over the whole thing. See Figure 4-17. Twin lead has two conductors spaced some distance apart, with plastic insulation between them. Some twin lead, called ladder line or open-wire feed line, is made from bare wires with plastic spacers placed every few inches to maintain the separation between the two conductors. See Figure 4-18. The distance between the conductors determines the characteristic impedance of the feed line. For 300-ohm line the conductors are about ½ inch apart. If the spacing is closer to 1 inch, the characteristic impedance is about 450 ohms.

Coaxial cable is an "unbalanced" feed line—the shield braid is connected to ground, while the center conductor is not. Twin lead is a "balanced" feed line—neither conductor is connected to ground. Most antennas are inherently balanced devices. For example, equal currents and voltages flow in both halves of a dipole antenna. Balanced antennas should ideally be fed with balanced feed line. When a balanced antenna is fed with unbalanced line, a device called a *balun* can be used. The balun (short for BALanced to UNbalanced) is used to transform the unbalanced condition on the feed line to the balanced condition at the antenna.

The Transmatch, or Impedance-Matching Network

An ideal antenna would be a purely resistive load. A practical antenna, however, will usually have an impedance with either a capacitive or inductive component in addition to resistance. A *Transmatch* is used to cancel out the capacitive or inductive component in the antenna impedance. With a Transmatch you can match the impedance of your antenna system to the impedance of the transmitter output, normally 50 ohms. Nonresonant antennas, such as random-wire antennas, exhibit widely varying impedances. A Transmatch makes these antennas usable on almost any amateur band. You will learn more about impedance matching in Chapter 5.

The Antenna Noise Bridge

An *antenna noise bridge* is a device that allows you to measure the impedance of antennas and other electrical circuits. The bridge produces a wide-band noise signal that is applied to the circuit under test. A receiver is also connected to the bridge and tuned to the desired operating frequency. You then adjust the bridge controls until the noise heard in the receiver is nulled (decreased to a minimum). By noting the settings of the controls on the noise bridge, and then calculating the reactance of the indicated capacitance or inductance, you can determine the impedance of the unknown circuit.

Wattmeters and Reflectometers

Other useful metering devices are the *wattmeter* and the *reflectometer* (SWR meter). A wattmeter is a device that is placed in the transmission line to measure the power (in watts) coming out of a transmitter. Wattmeters are designed to operate at a certain line impedance, normally 50 ohms. When using a wattmeter, make sure that the feed line impedance is the same as the design impedance of the watt-meter—otherwise, any measurements will be inaccurate. For most accurate measurement, the wattmeter should be connected directly at the transmitter antenna jack.

All wattmeters must contain some type of detection circuitry to enable you to measure power. The accuracy and upper frequency range of the wattmeter is limited by the detection device. Stray capacitance and coupling within the detection circuits can lead to a loss of sensitivity or accuracy as the operating frequency is increased.

A reflectometer is a device used to measure something called standing-wave ratio, or SWR. SWR is a measure of the relationship between the amount of power traveling to the antenna and the power reflected back to the transmitter by impedance mismatches in the antenna system. The energy going from the transmitter to the antenna is represented by one voltage (called forward or incident), and the energy reflected by the antenna is represented by the other (called reflected). A bridge circuit separates the incident and reflected voltages for measurement purposes. This is sufficient for determining SWR. Bridges designed to measure SWR are called reflectometers or SWR meters.

A reflectometer is placed in the transmission line between the transmitter and antenna. In the most common amateur application, the reflectometer is connected between the transmitter and Transmatch, and used to indicate minimum reflected power as the Transmatch is adjusted. This indicates when the antenna system (including feed line) is matched to the transmitter output impedance.

If you want to measure the impedance match between an antenna and the feed line, you should place the SWR meter right at the antenna feed point. This is because if you put the meter at the transmitter end, losses in the feed line make SWR readings look lower than they really are. More information on antennas and feed lines will be covered in Chapter 9.

Another device that can be used to measure SWR is called a *directional wattmeter*. This type of wattmeter either has one meter that reads forward power and another meter to read reflected power, or a single meter that can be switched to read either forward or reflected power.

[If you are studying for an Element 3A (Technician) exam, turn to Chapter 10 and study the following questions: 3D-5.2, 3D-11.1 through 3D-11.5, 3D-14.1 through 3D-14.5, 3D-14.8 and questions 3D-18.1 through 3D-18.3.

If you are studying for an Element 3B (General) exam, study questions 3D-14.6 and 3D-14.7 in Chapter 11.]

RADIO-FREQUENCY INTERFERENCE (RFI)

At one time or another, most amateurs have suffered the wrath of neighbors or family as their transmission was intercepted by a radio or television set. If you've never suffered problems like these, consider yourself lucky. An important means of survival in this age of electric and electronic devices is the ability to recognize and cure *radio-frequency interference*.

Even if your transmitter puts out no harmonics or other spurious signals, interference is still an unfortunate possibility. Your RF signal can result in audio interference to another electronic device, such as a radio, television, stereo, electronic organ or intercom. For the purpose of this discussion, we'll call these audio devices.

RFI to audio devices is usually caused by *audio rectification*. Audio rectification happens when the RF signal gets into the audio device and is rectified (detected). The resulting audio signal is then amplified by the audio device, resulting in interference.

When the amateur operator is using double-sideband AM voice (emission A3E), the interference to the electronic device will be heard like any normal signal applied to the amplifier. The amateur's voice will be heard through the electronic equipment, and may be only slightly distorted. Audio rectification from an amateur SSB transmitter (emission J3E) will sound like an unintelligible, garbled noise. If the amateur is using an FM transmitter (emission F3E), no sound will usually be heard: Instead, a decrease in volume will occur when the transmitter is on. Clicks may be heard when the transmitter is keyed on and off. Similar clicks will be heard if the interference is from a CW transmitter (emission A1A).

The first step in curing audio rectification involves finding out how the RF energy is entering the device, and "closing the door" with bypassing and filtering to keep it out. Several entry points are possible: the ac-line cord, the antenna connection (if any), speaker wires or interconnecting cables (in a component stereo system). Another possibility is direct radiation into the circuitry of the audio device itself (usually through a wooden or plastic case).

Speaker wires and the ac power cord on an audio device form good entry points for RF, especially for signals on the higher amateur bands. On the higher bands, these lines can become a sizable fraction of a wavelength, and act as an antenna. A simple ac line filter, shown in Figure 4-19, can be built from readily available parts and installed at the audio device. The preferred location for the filter is inside the audio device, where the ac line enters.

If RF is entering through external speaker wires, simply bypassing the wires to the chassis at the speaker terminals is usually effective. This technique is shown in Figure 4-20. In stubborn cases, it might be necessary to install the filter shown in Figure 4-21.

Installation of any of the devices mentioned in this section should be performed only by a qualified repair technician. If you install a filter in your neighbor's equipment, you may be blamed for any problems that occur with the equipment later, whether they relate to the work you did or not!

Another common victim of audio rectification is the telephone. If you or

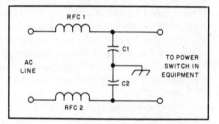

Figure 4-19—A method of ac-line filtering.
C1,C2—0.01-μF disc ceramic, 1.2-kV capacitors
RFC1, RFC2—24 turns number 18 AWG enameled wire, close spaced and wound on a ¼-inch-diameter form (such as a pencil).

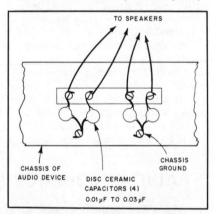

Figure 4-20—When speaker leads are bypassed, the disc capacitors should be mounted directly between the speaker terminals and the chassis, keeping the capacitor leads as short as possible.

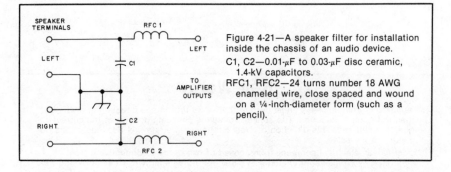

Figure 4-21—A speaker filter for installation inside the chassis of an audio device. C1, C2—0.01-μF to 0.03-μF disc ceramic, 1.4-kV capacitors. RFC1, RFC2—24 turns number 18 AWG enameled wire, close spaced and wound on a ¼-inch-diameter form (such as a pencil).

your neighbor are experiencing telephone interference, contact your local telephone company. They have inductors and other filter components to help eliminate the problem. With so many consumer-owned electronic telephones, and customers maintaining the phone wires inside their own homes, you may have some difficulty getting the phone company to help with the problem. Contacting them to ask for help is the best course of action, however.

Make sure your station equipment is operating properly, and that it does not cause interference to audio equipment in your own home. Your transmitter should be effectively grounded to a rod with the point driven into the ground at least 8 feet. The ground lead should be at least no. 10 wire or copper ribbon. Just remember, the greater the surface area of the ground lead, the more effective it will be. Also, keep the ground lead as short as possible.

Remember that amateurs may use only the amount of power necessary to establish the desired communication. Operating with excessive power is more likely to cause audio-rectification problems than with low power.

[If you are studying for an Element 3B (General) exam, turn to Chapter 11 and study the questions that begin with numbers 3D-10.]

Key Words

Alternating current (ac)—electrical current that flows first in one direction in a wire and then in the other direction. The applied voltage is changing polarity as the current direction changes. This direction reversal continues at a rate that depends on the frequency of the ac.

Capacitor—an electrical component composed of two or more conductive plates separated by an insulating material. A capacitor stores energy in an electrostatic field.

Coefficient of coupling—a measure of the amount of energy that will be transferred from one coil to another

Coil—a conductor wound into a series of loops

Core—the material in the center of a coil. The material used for the core affects the inductance value of the coil.

Current—a flow of electrons in an electrical circuit

Decibel (dB)—the smallest change in sound level that can be detected by the human ear. Power gains and losses are also expressed in decibels.

Dielectric—the insulating material used between plates in a capacitor

Direct current (dc)—electrical current that flows in one direction only

Effective voltage—the value of a dc voltage that will heat a resistive component to the same temperature as the ac voltage that is being measured

Electromotive force (EMF)—the force or pressure that pushes a current through a circuit

Farad—the basic unit of capacitance

Henry—the basic unit of inductance

Impedance—a term used to describe a combination of reactance and resistance in a circuit

Inductor—an electrical component usually composed of a coil of wire wound on a central core. An inductor stores energy in a magnetic field.

Mutual inductance—the ability of one coil to induce a voltage in another coil. Any two coils positioned so that the magnetic flux from one coil cuts through the turns of wire in the other are said to have mutual inductance.

Ohm's Law—a basic law of electronics, it gives a relationship between voltage, resistance and current ($E = IR$)

Ohm—the basic unit of resistance

Parallel circuit—an electrical circuit in which the electrons follow more than one path in going from the negative supply terminal to the positive terminal

Power—the rate at which energy is consumed. In an electric circuit, power is found by multiplying the voltage applied to the circuit by the current through the circuit.

Reactance—the opposition to current flow offered by a capacitor or inductor in an ac circuit

Resistance—the ability to oppose an electrical current

Root-mean-square (RMS) voltage—another name for effective voltage. The term refers to the method of calculating the value.

Series circuit—an electrical circuit in which all the electrons must flow through every part of the circuit. There is only one path for the electrons to follow.

Transformer—two coils with mutual inductance used to change the voltage level of an ac power source to one more suited for a particular circuit

Voltage—the EMF or pressure that causes electrons to move through an electrical circuit

Chapter 5

Electrical Principles

T o pass your FCC Element 2 written test and receive a Novice license, you had to learn some very basic radio theory. For your Element 3 test, you will need to understand a few more electrical principles. In addition to the basic concepts of resistance, capacitance and inductance, this chapter will introduce impedance and reactance.

Ohm's Law is an important tool for analyzing circuits; we'll look at some applications of this basic electronics principle. We will introduce two other useful analytical tools, Kirchhoff's voltage and current laws, and show you how to use them to analyze parallel and series circuits. We'll look at methods for finding the total value of parallel and series combinations of resistors, capacitors and inductors, and how the ratio of the primary and secondary turns on a transformer winding affects its operation.

Be sure to turn to Chapter 10 and study the appropriate FCC questions when you are directed to do so. This will show you if you are progressing smoothly, or if you need to do a little extra studying. We cannot tell you everything about a particular topic in one chapter—you may want to refer to some other reference books for additional information. *The ARRL Handbook for the Radio Amateur* is a good place to start, and *Understanding Amateur Radio,* also published by the ARRL, is another useful reference.

E = VOLTAGE

Electrons need a push to get them moving; we call the force that pushes electrons through a circuit *electromotive force (EMF)*. EMF is similar to water pressure in a pipe. The more pressure, the more water that flows through the pipe. The greater the EMF, the greater the flow of electrons in a circuit.

EMF is measured in volts, so it is usually called *voltage*. A voltage-measuring instrument is called a voltmeter. The more voltage applied to a circuit, the more electrons will flow through the circuit.

I = CURRENT

The word *current* comes to us from a Latin word meaning "to run" and it always implies movement. When someone speaks of the current in a river, we think of a flow of water; a current of air can be a light breeze or a hurricane; and a current of electricity is a flow of electrons.

Current is indicated in equations and diagrams by the letter I (from the French word *intensité*). We measure current in amperes, and the measuring device is called an ammeter. An ampere is often too large a unit for convenient use, so we can measure current in milliamperes (one thousandth of an ampere) or microamperes (one millionth of an ampere). Amperes are abbreviated amp or A; milliamperes mA and microamperes μA.

In considering current, it is natural to think of a single, constant force causing the electrons to move. When this is so, the electrons always move in the same direction through a circuit made up of conductors connected together in a continuous loop. Such a current is called a *direct current,* abbreviated *dc.* This is the type of current furnished by batteries and by certain types of generators.

Alternating Current

It is also possible to have a voltage source that periodically reverses polarity. With this kind of voltage the current flows first in one direction and then in the opposite direction. Such a voltage is called an alternating voltage, and the current is called an *alternating current* (abbreviated *ac*). The reversals (alternations) may occur at any rate from less than one per second up to several billion or more per second.

The difference between alternating current and direct current is shown in Figure 5-1. In these graphs, the horizontal axis indicates time, increasing toward the right. The vertical axis represents the amplitude or strength of the current, increasing in either the up or down direction away from the horizontal axis. If the graph is above the horizontal axis, the current is flowing in one direction through the circuit (indicated by the + sign). If the graph is below the horizontal axis the current is flowing in the opposite direction (indicated by the − sign).

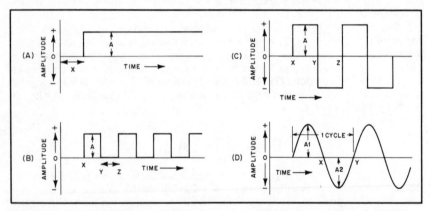

Figure 5-1—Four types of current are shown. A—direct current; B—intermittent (pulsating) direct current; C—square-wave alternating current; D—sine-wave alternating current.

In Figure 5-1A, assume that we close the circuit or make the path for the current complete at the time indicated by X. The current instantly takes the amplitude indicated by the height A. After that, the current continues at the same amplitude as time goes on. This is an ordinary direct current.

In Figure 5-1B, the current starts flowing with the amplitude A at time X, continues at that amplitude until time Y and then instantly ceases. After an interval from Y to Z the current begins to flow again and the same sort of stop-and-start performance is repeated. This is an intermittent direct current (also called a pulsating direct current). We could produce this type of current by alternately opening and closing a switch in the circuit. It is a direct current because the direction of the current does not change; the amplitude is always on the + side of the horizontal axis.

The waveform in Figure 5-1C is a little different. The current starts with amplitude A at time X just as the wave in Figure 5-1B, but at time Y the current instantly reverses direction. From time Y to time Z the current flows in the opposite direction and the voltage has an equal, but opposite polarity. At time Z the current reverses direction again. This is called a square-wave alternating current.

In Figure 5-1D the current starts at zero, smoothly increases in amplitude until it reaches the amplitude A1 while flowing in the + direction, then decreases smoothly until it drops to zero amplitude once more. At that time (X) the direction of the current reverses. As time goes on the amplitude increases, with the current now flowing in the − direction, until it reaches amplitude A2. Then the amplitude decreases until finally it returns to zero (at point Y) and the direction reverses once more. The type of smoothly varying alternating current shown in Figure 5-1D is known as a sine wave.

Looking at the sine wave, the first thing we notice is that the waveform reaches a maximum value (peak) at two points in the ac cycle—once during the positive part and once during the negative part. For the remainder of the cycle, the waveform is constantly varying between these two values. Another interesting characteristic of a sine wave is that the average dc voltage contained in the wave is zero. The positive half of the cycle is a "mirror image" of the negative side—for every positive value, there is a corresponding negative value, and the positive and negative values cancel each other out. This doesn't mean that a sine wave contains no energy, however—it means only that the average dc potential is zero.

So how can we find a value for ac voltage and current? The wave is constantly changing. If we could measure the voltage at one point on the wave at an instant of time, in the next instant the value would be different. There is no one value we can point to and say: "That's the value of the voltage."

We could use the peak value the wave reaches on its positive half-cycle. Or we could measure the voltage from the positive peak to the negative peak, and use this to specify the current or voltage. This value is called the peak-to-peak value. We can also use a measuring scale that lets us compare the ac voltage to an equivalent dc voltage.

If we pass dc through a resistor, the resistor will get warm. If we pass ac through the same resistor, it will also get warm. The ac voltage that will heat a resistor to the same temperature as 1-V dc is said to have an *effective voltage* of 1-V ac. If we pass 10-V dc through a resistor and measure its temperature, we can then find an effective ac voltage of 10-V ac that will heat the resistor to the same temperature.

Most ac voltages are specified by their effective values. Your wall outlets provide 117-V effective ac. If we plug a soldering iron into a source of 117-V dc, it will get just as hot as it will if we plug it into the wall outlet.

The effective voltage is also called the *root-mean-square (RMS)* value of the ac wave. The phrase root-mean-square describes the actual process of calculating the effective voltage. The method involves squaring the instantaneous value for a large number of points along the waveform and then finding the average of the squared values. We then take the square root of the average (the **root** of the average (**mean**) of the **squares**).

If we know the peak-to-peak value of an ac waveform, how do we calculate the RMS value? The RMS value and the peak value are related by a factor of the square root of two. For sine waves, the following relationships hold:

$$V_{peak} = V_{RMS} \times \sqrt{2} = V_{RMS} \times 1.414 \qquad \text{(Equation 5-1)}$$

$$V_{RMS} = \frac{V_{peak}}{\sqrt{2}} = \frac{V_{peak}}{1.414} = V_{peak} \times 0.707 \qquad \text{(Equation 5-2)}$$

Remember that the peak value is one-half the peak-to-peak value. Figure 5-2 shows the relationships between peak, peak-to-peak and RMS ac values.

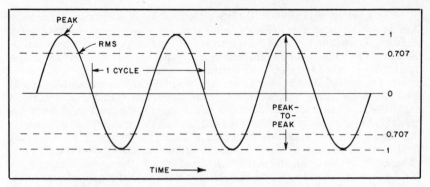

Figure 5-2—The RMS (effective) value of an ac waveform is 0.707 times the peak value. The peak value is one-half the peak-to-peak value.

For example, to find the peak-to-peak value of a 10-V RMS ac wave, we use Equation 5-1:

$$V_{peak} = V_{RMS} \times \sqrt{2}$$

$$V_{peak} = 10 \text{ V} \times 1.414 = 14.14 \text{ V}$$

Remember that the peak-to-peak voltage is peak voltage times 2, so:

$$V_{peak\text{-}to\text{-}peak} = 14.14 \text{ V} \times 2 = 28.28 \text{ V}$$

To find the RMS value of a 50-V peak-to-peak ac wave, we use Equation 5-2:

$$V_{peak} = \frac{V_{peak\text{-}to\text{-}peak}}{2} = \frac{50 \text{ V}}{2} = 25 \text{ V}$$

$$V_{RMS} = \frac{V_{peak}}{\sqrt{2}}$$

$$V_{RMS} = \frac{25 \text{ V}}{1.414} = 17.7 \text{ V}$$

[If you are studying for an Element 3B (General) exam, study questions 3E-16.1 through 3E-16.3 in Chapter 11.]

R = RESISTANCE

Resistance is something that opposes motion. All materials have electrical resistance; there is always some opposition to the flow of electrons through the material, although the resistance of different materials varies widely. Some materials have very high resistance; we call these materials good insulators, or poor conductors. Other materials have very low resistance; we call these materials poor insulators, or good conductors. The basic unit of resistance is the ohm, and a resistance-measuring instrument is called an ohmmeter. The symbol for an ohm is Ω, the Greek letter capital omega.

Between the extremes of a very good insulator and a very good conductor, we can make devices with a range of opposition to the flow of electrons in a circuit. We call such devices resistors. If we use the analogy of water flowing in a pipe to help us understand electron flow, these materials act like a pipe with a sponge in it. Some water will still flow in the pipe, but the flow will be reduced. The primary function of a resistor is to limit electron flow (current). The greater the resistance, the greater the reduction in current. If we want to control the current reduction over a certain range, we can use a variable resistor. A variable resistor, as its name implies, can be made to change resistance over a range. We can then control the flow of electrons from a maximum to a minimum over the variable resistor's range.

This current reduction has a price, however; the energy lost by the electrons as they flow through a resistor must be dissipated in some way. The energy is converted into heat, and the resistor gets warm. If the current is higher than the resistor can handle, the resistor will get very hot and may even be destroyed.

[If you are studying for an Element 3A (Technician) exam, turn to Chapter 10 and study questions 3E-2.1 through 3E-2.4.]

Resistors in Series and Parallel

Sometimes it's necessary to calculate the total resistance of resistors connected in a *series circuit* (end to end like a string of sausages) or a *parallel circuit* (side by side like a picket fence). One common occurrence is when a certain amount of resistance is needed somewhere in a circuit and there is no standard resistor value that will give the necessary resistance. In some cases you may not have a certain value on hand, but by combining resistors in parallel or series you can obtain the desired value.

When resistors are connected in series, as in Figure 5-3, the total resistance is simply the sum of all of the resistors. If we use the analogy of a sponge in a water pipe again, connecting resistors in series is like putting several sponges in the same pipe. The total resistance to the water would be the sum of all the individual resistances added up. So resistors in series add:

$$R_{TOTAL} = R_1 + R_2 + R_3 + \ldots + R_n \qquad \text{(Equation 5-3)}$$

Note that the total resistance of a string of resistors in series will always be greater than the individual resistors in the string. All the circuit current flows through each resistor in a series circuit.

In a parallel circuit, things are a bit different. When two or more resistors are connected in parallel, there is more than one path for the current in the circuit. See Figure 5-4. This is like connecting another water pipe of the same diameter into our

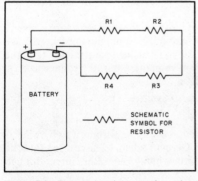

Figure 5-3—The total resistance of a string of series-connected resistors is the sum of all the individual resistances.

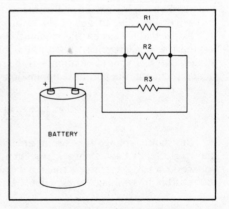

Figure 5-4—Parallel-connected resistors.

water-pipe circuit. When there is more than one path, more water can flow over a given time. With more than one resistor, more electrons can flow, and this means there is a greater current.

The formula for calculating the total resistance of resistors connected in parallel is:

$$R_{TOTAL} = \cfrac{1}{\cfrac{1}{R_1} + \cfrac{1}{R_2} + \cfrac{1}{R_3} + \ldots + \cfrac{1}{R_n}} \qquad \text{(Equation 5-4)}$$

where n is the total number of resistors. For example, if we connect three 100-ohm resistors in parallel, their total resistance is:

$$R_{TOTAL} = \cfrac{1}{\cfrac{1}{100} + \cfrac{1}{100} + \cfrac{1}{100}}$$

$$R_{TOTAL} = \cfrac{1}{\cfrac{3}{100}} = \frac{1}{0.03} = 33.33 \text{ ohms}$$

If we connect a 25-ohm, a 100-ohm and a 10-ohm resistor, our equation becomes:

$$R_{TOTAL} = \cfrac{1}{\cfrac{1}{25} + \cfrac{1}{100} + \cfrac{1}{10}}$$

$$R_{TOTAL} = \frac{1}{0.04 + 0.01 + 0.1} = \frac{1}{0.15} = 6.67 \text{ ohms}$$

To calculate the total resistance of two resistors in parallel, Equation 5-4 reduces to the "product over sum" formula. Divide the product of the two resistances by their sum:

$$R_{TOTAL} = \frac{R_1 \times R_2}{R_1 + R_2} \qquad \text{(Equation 5-5)}$$

If we connect two 50-ohm resistors in parallel, the total resistance would be:

$$R_{TOTAL} = \frac{50 \times 50}{50 + 50} = \frac{2500}{100} = 25 \text{ ohms}$$

From this we can see that the total resistance of two equal resistors connected in parallel is one-half the value of one of the resistors.

Notice that for any parallel combination of resistors, the total resistance is always less than the smallest value of the parallel combination. You can use this fact to make a quick check of your calculations—if the result you calculate isn't smaller than the smallest value in the parallel combination, you've made a mistake somewhere!

OHM'S LAW

Resistance, voltage and current are related in a very important way by what is known as Ohm's Law. Georg Simon Ohm, a German scientist, discovered that the voltage drop across a resistor is equal to the resistance multiplied by the current through the resistor. In symbolic terms, Ohm's Law can be written:

$$E = I \times R \qquad \text{(Equation 5-6)}$$

By simply rearranging the terms, we can show that the current is equal to the voltage drop divided by the resistance:

$$I = \frac{E}{R}$$

(Equation 5-7)

and that the resistance equals the voltage drop divided by the current:

$$R = \frac{E}{I}$$

(Equation 5-8)

Ohm's Law is an important mathematical relationship; you should understand and memorize it. A simple way to remember the Ohm's Law equations is shown in Figure 5-5. If we draw the "Ohm's Law circle" shown, we can find the equation for any of the three quantities in the relationship.

To use the circle, just cover the letter you want to solve the equation for. If the two remaining letters are across from each other, you must multiply them. If the two letters are arranged one on top of the other you must divide the top letter by the bottom. For example, if you cover I, you have "E over R" (Equation 5-7). Cover E, and you have "I times R" (Equation 5-6).

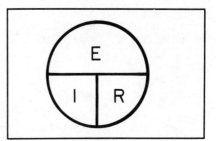

Figure 5-5—The "Ohm's Law Circle." Cover the letter representing the unknown quantity to find a formula to calculate that quantity. For example, if you cover the I, you are left with E / R.

When you know any of the two quantities in Ohm's Law, you can always find the third. See Figure 5-6. Suppose you have a circuit where the resistance is 10 ohms

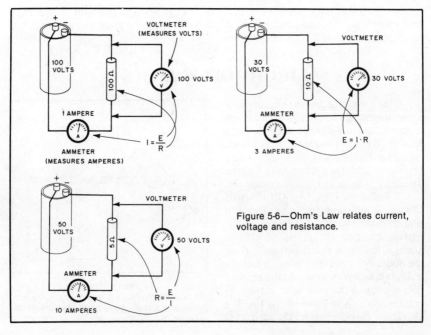

Figure 5-6—Ohm's Law relates current, voltage and resistance.

and the current is 3 amperes. To find the voltage, multiply the resistance times the current:

$$E = I \times R$$

$$E = 3 \text{ amps} \times 10 \text{ ohms}$$

$$E = 30 \text{ volts}$$

In a circuit with 100 volts applied to a 100-ohm resistor, you would use Equation 5-2 to find the current. Substituting the values for our problem into the formula, we have:

$$I = \frac{E}{R}$$
$$I = \frac{100 \text{ volts}}{100 \text{ ohms}} = 1 \text{ amp}$$

In a circuit with 50 volts applied and 10 amps of current flowing, we use Equation 5-3 to find the resistance:

$$R = \frac{E}{I}$$
$$R = \frac{50 \text{ volts}}{10 \text{ amps}} = 5 \text{ ohms}$$

To check our work we can always work the problem back to find one of the given quantities. In the last problem, using $E = I \times R$:

$$E = 10 \text{ amps} \times 5 \text{ ohms} = 50 \text{ volts}$$

And we prove that our math was correct! You won't go wrong if you write the equation you want to use first, and then substitute the values you are given and solve for the unknown, as shown in these examples.

KIRCHHOFF'S LAWS

When we analyze a circuit, it is sometimes useful to calculate the current through, and the voltage drop across a resistor. We have two electrical principles, called Kirchhoff's laws, that make it easy for us to find the current and voltage drop.

Kirchhoff's First Law (Kirchhoff's Current Law) states that when a circuit branches into two or more directions, the current entering the branch point, or node, is the same as the current leaving the node. We can use this law to find an unknown quantity in a parallel circuit.

We know that the total current I_T entering point A in Figure 5-7 is 1 A. We also know that the current through R1 is

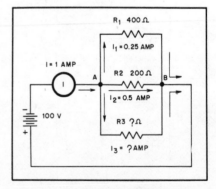

Figure 5-7—Kirchhoff's First Law: The sum of the currents flowing into a junction point (node) in a circuit must equal the sum of the currents leaving the node.

0.25 A and that the current through R2 is 0.5 A. The current through R3 must therefore be 0.25 A since:

$$I_1 + I_2 + I_3 = I_T = 1 \text{ amp}$$

$$0.25 \text{ amp} + 0.5 \text{ amp} + I_3 = 1 \text{ amp}$$

$$I_3 = 1 \text{ amp} - 0.25 \text{ amp} - 0.5 \text{ amp} = 0.25 \text{ amp}$$

Once we know the current through R_3 we can find its resistance. Since we know that 100 V is dropped from point A to B, we can use Equation 5-8 to find the resistance:

$$R = \frac{E}{I}$$

$$R = \frac{100 \text{ volts}}{0.25 \text{ amps}} = 400 \text{ ohms}$$

Kirchhoff's Second Law (Kirchhoff's Voltage Law) tells us that the sum of the voltage drops must equal the sum of the voltage rises around any closed loop in a circuit (such as in a series circuit). So in the series circuit in Figure 5-8, the sum of the voltage drops across the resistors must equal the applied voltage. The voltage drop across each resistor can be found by multiplying the current through the resistor by the resistance. (This is a simple Ohm's Law relationship, $E = I \times R$.) Since all the current in the circuit flows through each resistor, we can find the value of current by dividing the applied voltage (120 V) by the total

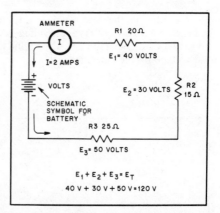

Figure 5-8—Kirchhoff's Second Law: The sum of the individual voltage drops in a closed loop (such as a series circuit) equals the sum of the voltage rises.

resistance of the circuit, 60 ohms. From this we find that 2 amps flows through each resistor. We can then calculate the voltage drops.

$$E_{R1} = 2 \text{ A} \times 20 \text{ ohms} = 40 \text{ V}$$
$$E_{R2} = 2 \text{ A} \times 15 \text{ ohms} = 30 \text{ V}$$
$$E_{R3} = 2 \text{ A} \times 25 \text{ ohms} = \underline{50 \text{ V}}$$
Total voltage drop = 120 V

Notice that the larger the resistance, the larger the voltage drop across that resistor.

[If you are studying for an Element 3A (Technician) exam, turn to Chapter 10 and study question 3E-12.2 and questions 3E-14.8 through 3E-14.11.

If you are studying for an Element 3B (General) exam, study the following questions in Chapter 11: 3E-12.1, 3E-12.3, 3E-14.1, 3E-14.2 and 3E-14.5.]

P = POWER

Suppose we want to know how fast energy is being consumed in a circuit. If you want to compare how bright two different light bulbs will be, or how much it will cost to run a new freezer for a month, you will have to know how fast they use electrical energy. *Power* is defined as the time rate of energy consumption. The basic unit for measuring power is the watt, named for James Watt, the inventor of the steam engine.

Since current is a measure of how fast electrons are flowing through a circuit, and voltage is a measure of the force needed to move the electrons, there is a very simple way to calculate power. We use the equation:

$$P = I \times E \qquad \text{(Equation 5-9)}$$

where
  P = power, measured in watts
  E = the EMF in volts
  I = the current in amperes

To calculate the power in a circuit, multiply volts times amperes. For example, in Figure 5-9 a 12-V battery is providing 3 A of current to a light bulb that is operating normally. Using Equation 5-9, we can find the power rating for the bulb by multiplying the voltage by the current:

$$P = I \times E = 3 \text{ A} \times 12 \text{ V} = 36 \text{ watts}$$

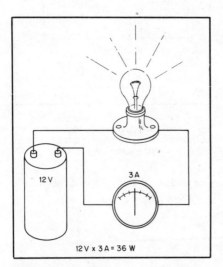

12 V × 3 A = 36 W

Figure 5-9—Applied voltage multiplied by current equals power.

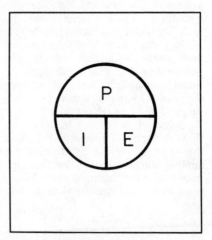

Figure 5-10—The "Power Circle." Cover the letter representing the unknown quantity to find a formula to calculate that quantity. If you cover the P, you are left with I × E.

If you know any two of the three quantities in Equation 5-9, you can find the unknown third quantity, just as you can find the unknown in an Ohm's Law problem. We can draw a "power circle" like the Ohm's Law circle, as shown in Figure 5-10. Cover the unknown quantity to find the equation to use.

$$I = \frac{P}{E} \qquad \text{(Equation 5-10)}$$

$$E = \frac{P}{I} \qquad \text{(Equation 5-11)}$$

If 10 V is applied to a circuit consuming 20 W, how much current is flowing in the circuit? Use Equation 5-10 to find the current:

$$I = \frac{P}{E}$$

$$I = \frac{20 \text{ W}}{10 \text{ V}} = 2 \text{ amps}$$

If a 5-A current flows in a 60-W circuit, how much voltage is applied to the circuit? Equation 5-11 will help us solve this problem:

$$E = \frac{P}{I}$$

$$E = \frac{60 \text{ W}}{5 \text{ A}} = 12 \text{ volts}$$

If we apply 500 volts to a 200,000-ohm resistor, how much power will the resistor have to dissipate? First, we use Ohm's Law to find the current in the circuit:

$$I = \frac{E}{R} = \frac{500 \text{ volts}}{200,000 \text{ ohms}} = 0.0025 \text{ amp}$$

Then, we can use Equation 5-9 to find the power:

$$P = I \times E = 0.0025 \text{ amp} \times 500 \text{ volts} = 1.25 \text{ watts}$$

What if we want to know the power dissipated in a 1000-ohm resistor that has 0.3 amp flowing through it? To use Equation 5-9 we need to know the applied voltage—but we can find that with Ohm's Law:

$$E = I \times R = 0.3 \text{ amp} \times 1000 \text{ ohms} = 300 \text{ volts}$$

Now we can use Equation 5-9:

$$P = I \times E = 300 \text{ volts} \times 0.3 \text{ amp} = 90 \text{ watts}$$

[If you are studying for an Element 3A (Technician) exam, turn to Chapter 10 and study questions 3E-11.1 through 3E-11.5 and question 3E-13.4.

If you are studying for an Element 3B (General) exam, study questions 3E-13.1 through 3E-13.3 in Chapter 11.]

Power Gain, Power Loss and the Decibel

Power gains and losses in electrical circuits are often spoken of in terms of *decibels* (one-tenth of a bel). The bel was named for Alexander Graham Bell, and was first used in early audio telephone work to describe a change in loudness levels. The decibel is defined as the change in sound power level that is just perceptible to the human ear. In mathematical terms power gain or loss in decibels is given by the equation:

$$dB = 10 \times \log_{10} \left(\frac{P_2}{P_1} \right) \qquad \text{(Equation 5-12)}$$

where
P_1 = reference power
P_2 = power being compared to the reference value

The $\log_{10}$ in this equation means "the logarithm to the base 10," also called the common logarithm. The common logarithm of a number is defined as the power that you must raise 10 to in order to obtain that number. If $X = 10^A$, then A is the $\log_{10}$ of X. Your scientific calculator probably has keys labeled "log" and "y^x" to help you with these calculations.

The $\log_{10}$ of 10 is 1, because 10 raised to the first power is 10 ($10^1 = 10$). The $\log_{10}$ of 100 is 2 ($10^2 = 100$); the $\log_{10}$ of 10,000 is 4, and so on. You may notice from these examples that it is very easy to find the common logarithm of a number consisting of a one and any number of zeros. Simply count the zeros to find the logarithm! The logarithm gives us the answer to the question "10 raised to what power will give us this number?"

Let's look at some sample problems using the power gain and loss relationship. If we increase our transmitter power level from 100 watts to 1000 watts, this would be a power gain of:

$$dB = 10 \times \log_{10}\left(\frac{P_2}{P_1}\right)$$

$$dB = 10 \times \log_{10}\left(\frac{1000 \text{ W}}{100 \text{ W}}\right)$$

$$dB = 10 \times \log_{10}(10)$$

$$dB = 10 \times 1 = 10 \text{ dB}$$

A power decrease from 1500 watts to 15 watts would result in a loss of 20 dB:

$$dB = 10 \times \log_{10}\left(\frac{P_2}{P_1}\right)$$

$$dB = 10 \times \log_{10}\left(\frac{15 \text{ W}}{1500 \text{ W}}\right)$$

$$dB = 10 \times \log_{10}(0.01)$$

$$dB = 10 \times -2 = -20 \text{ dB}$$

Notice that the decrease from 1500 watts to 15 watts can be thought of as one drop of 10 dB from 1500 W to 150 W, and another 10 dB drop from 150 W to 15 W, so the total power loss is $-10 \text{ dB} + -10 \text{ dB} = -20 \text{ dB}$. This is a very powerful property of logarithms. For example, if we double our transmitter power from 10 watts to 20 watts, we have a 3-dB gain:

$$dB = 10 \times \log_{10}\left(\frac{20 \text{ W}}{10 \text{ W}}\right)$$

$$dB = 10 \times \log_{10}(2)$$

$$dB = 10 \times 0.301 = \text{approximately 3-dB gain}$$

If we then double the power again to 40 watts, we have another 3-dB gain:

$$dB = 10 \times \log_{10}\left(\frac{40 \text{ W}}{20 \text{ W}}\right)$$

$$dB = 10 \times \log_{10}(2)$$

dB = 10 × 0.301 = approximately 3-dB gain

So if we multiply our power level four times from 10 watts to 40 watts, we would have a total power gain of 3 dB + 3 dB = 6 dB. Checking our calculations:

$$dB = 10 \times \log_{10} \left(\frac{40 \text{ W}}{10 \text{ W}} \right)$$

$$dB = 10 \times \log_{10} (4)$$

$$dB = 10 \times 0.602 = \text{approximately 6-dB gain}$$

Power gains and losses expressed in decibels can be added (or subtracted) to find the total system gain or loss. You will learn more about the applications of this principle when you study for your Advanced class license. For now, you just need to know the basic ideas presented here.

[If you are studying for an Element 3B (General) exam, study questions 3E-10.1 through 3E-10.9 in Chapter 11.]

CAPACITANCE

A simple capacitor is formed by separating two conductive plates with an insulating material, or *dielectric*. If we connect one plate to the positive terminal of a voltage source and the other plate to the negative terminal, we can build up a surplus of electrons on one plate, as shown in Figure 5-11. At some point, the voltage across

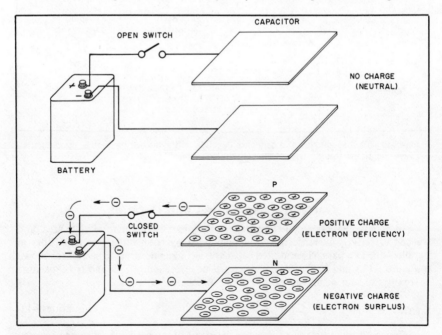

Figure 5-11—When a voltage is applied to a capacitor, an electron surplus (negative charge) builds up on one plate, while an electron deficiency forms on the other plate to produce a positive charge.

the capacitor will equal the applied voltage, and the capacitor is said to be charged. If we then connect a load to the capacitor, it will discharge through the load, releasing stored energy. The basic property of a capacitor is this ability to store a charge in an electric field.

The basic unit of capacitance is the farad, named for Michael Faraday. The farad is usually too large a unit for convenient measure, so we use microfarads (10^{-6}), abbreviated μF, or picofarads (10^{-12}), abbreviated pF.

Factors Determining Capacitance

The capacitance value of a capacitor is determined by three factors: the area of the plate surfaces, the distance between the plates and the type of insulating material (dielectric) used between the plates. See Figure 5-12. Increasing the area of the plates, reducing the spacing between them or using a better insulator as the dielectric will increase the capacitance. The effects of these three factors will be covered in more detail in Chapter 6.

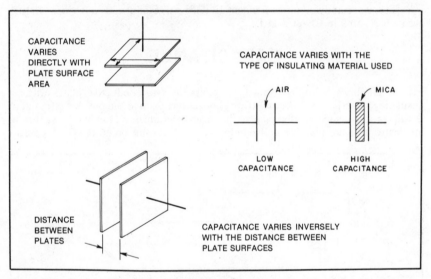

Figure 5-12—The capacitance of a capacitor depends on the area of the plates, the distance between the plates and the type of dielectric material used.

Capacitors in Series and Parallel

We can increase the value of capacitance by increasing the total plate area, and we can effectively increase the total plate area by connecting two capacitors in parallel—the two parallel-connected capacitors act like one larger capacitor, as shown in Figure 5-13A and B. The total capacitance of several parallel-connected capacitors is simply the sum of all the values added together:

$$C_{TOTAL} = C_1 + C_2 + C_3 + \ldots + C_n \qquad \text{(Equation 5-13)}$$

Connecting capacitors in series has the effect of increasing the distance between the plates, thereby reducing the total capacitance, as shown in Figure 5-13C and D. For capacitors in series, we use the familiar reciprocal formula:

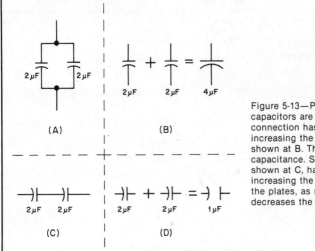

Figure 5-13—Parallel-connected capacitors are shown at A. This connection has the effect of increasing the total plate area, as shown at B. This increases the capacitance. Series connection, shown at C, has the effect of increasing the spacing between the plates, as shown at D. This decreases the capacitance.

$$C_{TOTAL} = \cfrac{1}{\cfrac{1}{C_1} + \cfrac{1}{C_2} + \cfrac{1}{C_3} + \ldots + \cfrac{1}{C_n}}$$

(Equation 5-14)

For only two capacitors in series, Equation 5-14 reduces to the product over sum formula:

$$C_{TOTAL} = \frac{C_1 \times C_2}{C_1 + C_2}$$

(Equation 5-15)

As we discovered using this formula for two equal resistors, for two equal capacitors connected in series the total capacitance will be half the value of one of the capacitors.

[If you are studying for an Element 3A (Technician) exam, turn to Chapter 10 and study questions in subelement 3E that begin 3E-5 and 3E-8.

If you are studying for an Element 3B (General) exam, study questions 3E-14.4 and 3E-14.7 in Chapter 11.]

INDUCTORS

The motion of electrons produces magnetism—every electric current creates a magnetic field around the wire in which it flows. Like an invisible tube, the magnetic field extends outward in concentric circles around the conductor. See Figure 5-14A. The field spreads outward as the current flows and collapses back into the conductor when the current stops. The field increases in strength when the current increases and decreases in strength as the current decreases.

When the conductor is a straight piece of wire, the force produced by this magnetic field is usually negligible compared to the force if the same wire is formed into a coil. In coils, the magnetic field around each turn also affects the other turns. Together the combined forces produce one large magnetic field, as shown in Figure 5-14B. Much of the energy in the magnetic field is concentrated in the material in the center of the coil (the *core*). Most practical *inductors* consist of a length of wire wound on a core.

In much the same way that a capacitor stores energy in an electrostatic field, an inductor stores energy in a magnetic field. When a dc voltage is first applied to

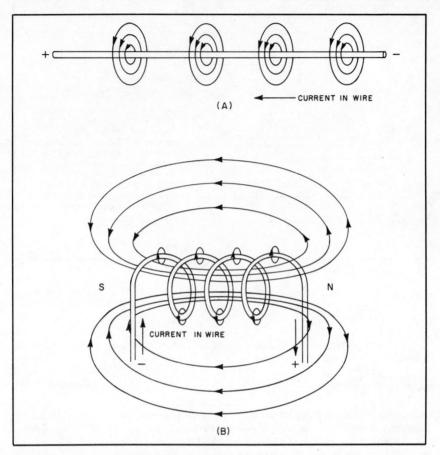

Figure 5-14—A magnetic field surrounds a wire with a current flowing in it. If the wire is formed into a coil, the magnetic field becomes much stronger as the lines of force reinforce each other.

an inductor (as when a switch is closed), with no current flowing through the circuit, the inductor will oppose the current. In fact, this is a basic property of inductors. They oppose any change in the current flowing through them. A voltage is induced in the coil that opposes the applied voltage, and tries to prevent the current from changing. This induced voltage is called a back EMF. Gradually a current will be produced by the applied voltage. (Actually this current is produced very quickly, but it does not start instantly when the switch is closed, as you might expect.) The "final" current that flows through the inductor is limited only by any resistance that might be in the circuit. There is very little resistance in the wire of most coils, so the current will be quite large if there is no other resistance.

In the process of getting this current to flow, energy is stored in the form of the magnetic field that is produced around the coil. When the applied current is shut off, the magnetic field collapses and returns the energy to the circuit as a current flowing in the opposite direction. This current can be quite large for large values of inductance, and the reverse voltage can rise to many times the applied voltage. This can cause a spark to jump across switch or relay contacts when the circuit is broken to turn off the current. This effect is called "inductive kickback."

When an ac voltage is applied to an inductor, the current flowing through the inductor will reverse direction every half cycle. This means the current will be constantly changing, and the inductor will oppose this change. The energy stored in the magnetic field while the current is increasing during the first half cycle will be returned to the circuit as the current starts to decrease. Then a new magnetic field will be produced during the second half cycle (with the north and south poles of the field pointing in opposite directions from the way they were pointing during the first half cycle). The energy stored in that field will be returned to the circuit as the current again starts to decrease, and then a new magnetic field will be produced on the next half cycle. This process keeps repeating, as long as the ac voltage is applied to the inductor.

Factors That Determine Inductance

The amount of opposition to changes in current, the amount of energy stored in the magnetic field and the back EMF induced in the coil all depend on the inductance of the coil. The inductance of a coil, usually represented by the symbol L, depends on four things:

1) the type of material used for the core (permeability of the material), and its size and location in the coil;
2) the number of turns used to wind the coil;
3) the length of the coil (spacing of turns);
4) the diameter of the coil (cross-sectional area)

Changing any of these factors changes the inductance. See Figure 5-15.

The basic unit of inductance is the henry, named for the American physicist, Joseph Henry. Like the farad, the henry is usually too large a unit for practical measurement, so we use the millihenry (10^{-3}) abbreviated mH, or microhenry (10^{-6}) abbreviated μH.

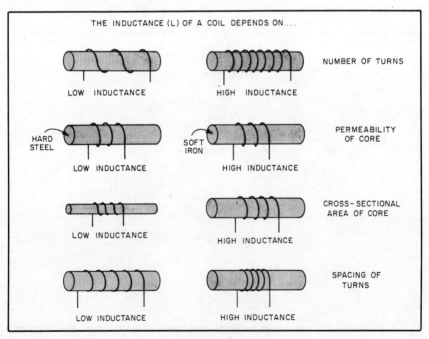

Figure 5-15—The value of an inductor depends on the material used in the core, the number of windings, and the length and diameter of the coil.

Inductors in Series and Parallel

In circuits, inductors combine like resistors. The total inductance of several inductors connected in series is the sum of all the inductors:

$$L_{TOTAL} = L_1 + L_2 + L_3 + \ldots + L_n \qquad \text{(Equation 5-16)}$$

where n is the total number of inductors.

For parallel-connected inductors:

$$L_{TOTAL} = \cfrac{1}{\cfrac{1}{L_1} + \cfrac{1}{L_2} + \cfrac{1}{L_3} + \ldots + \cfrac{1}{L_n}} \qquad \text{(Equation 5-17)}$$

You should recognize this equation by now, and realize that for two parallel-connected
inductors, Equation 5-17 reduces to:

$$L_{TOTAL} = \frac{L_1 \times L_2}{L_1 + L_2} \qquad \text{(Equation 5-18)}$$

Remember that for two equal components connected in parallel, the total value found
using this formula will be one-half the value of one of the components.

[If you are studying for an Element 3A (Technician) exam, turn to Chapter 10
and study questions in subelement 3E that begin 3E-4 and 3E-9.

If you are studying for an Element 3B (General) exam, study questions 3E-14.3
and 3E-14.6 in Chapter 11.]

Mutual Inductance

If two coils are arranged so that they line up as shown in Figure 5-16, a current

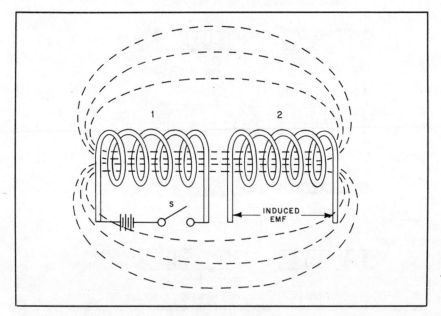

Figure 5-16—When the magnetic force produced by one coil cuts the turns of another coil, a
voltage will be induced in the second coil.

passed through coil 1 will produce a magnetic field which "cuts" coil 2. Consequently, a voltage will be induced in coil 2 whenever the field strength is changing. Any two coils positioned so that the flux from one coil cuts through the wires of the other coil have *mutual inductance*. We say the inductors are linked, or coupled.

If all of the flux lines set up by one coil cut all the turns of the other coil, the mutual inductance has its maximum possible value. If only a small part of the flux set up by one coil cuts the turns of the other, the mutual inductance is relatively small.

The ratio of actual mutual inductance to the maximum possible value that could theoretically be obtained with two given coils is called the *coefficient of coupling* between the two coils. This is frequently expressed as a percentage. The degree of coupling depends on the distance between the two coils and how they are positioned with respect to each other. See Figure 5-17. Maximum coupling occurs when the two coils have a common axis and are wound as close together as possible (one wound over the other). Coupling is least when the coils are far apart or are placed so that they are at right angles to each other. The maximum possible coefficient of coupling is approached only when the two coils are wound on a closed iron core.

Mutual inductance will affect the total inductance of coils connected in series or parallel. Equations 5-16, 5-17 and 5-18 are true only if the coils have no mutual inductance.

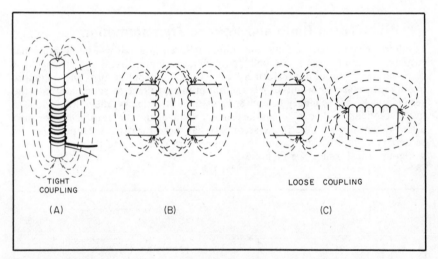

Figure 5-17—Maximum coupling between two coils will occur when the coils are wound directly over each other, as at A. Minimum coupling occurs when the coils are at right angles to each other, as at C.

TRANSFORMERS

When two coils are arranged so that a changing current in one induces a voltage in the other, the combination of windings is called a *transformer*. See Figure 5-18. Every transformer has a primary winding and a secondary winding. The primary winding is connected to the current source, and the secondary winding is connected to the load.

With a transformer, electrical energy can be transferred from one circuit to another without direct connection. In the process, the signal voltage can be readily changed from one level to another.

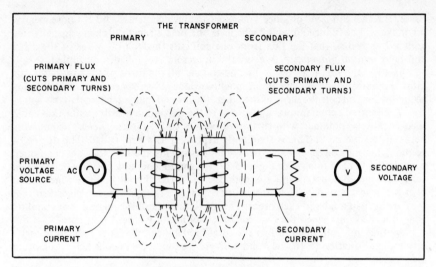

Figure 5-18—The transformer is an application of mutual inductance.

Turns Ratio and Voltage Transformation

For a given magnetic field, the voltage induced in a coil will be proportional to the number of turns in the coil. It follows, therefore, that if the primary and secondary windings of a transformer have the same number of turns and we ignore losses in the coils, we would expect the full primary voltage to be induced into the secondary. If the secondary has fewer turns than the primary, the voltage induced in the secondary will be less than the primary voltage, and the transformer is called a step-down transformer. If the secondary has more turns than the primary, the secondary voltage will be greater than the primary voltage, and the transformer is a step-up transformer. See Figure 5-19.

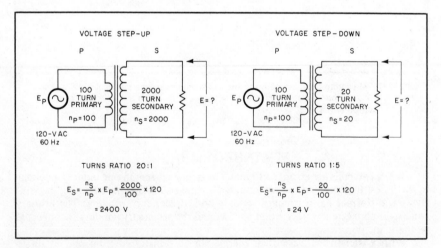

Figure 5-19—The amount of voltage transformation is controlled by the ratio of turns in the primary to turns in the secondary.

If you know the number of turns on the primary and secondary windings of a transformer and you know the applied voltage, it is easy to calculate the secondary voltage. We use the formula:

$$E_s = \frac{n_s}{n_p} \times E_p \hspace{4cm} \text{(Equation 5-18)}$$

where
E = voltage
n = number of turns
s = secondary
p = primary

For example, a transformer with 100 turns on the primary and 10 turns on the secondary with 117-V applied to the primary will produce how many volts at the secondary?

$$E_s = \frac{n_s}{n_p} \times E_p$$

$$E_s = \frac{10 \text{ turns}}{100 \text{ turns}} \times 117 \text{ V}$$

$$E_s = 0.1 \times 117 = 11.7 \text{ V}$$

If the primary has 20 turns and the secondary has 100 turns what secondary voltage will be produced with 10 V on the primary?

$$E_s = \frac{n_s}{n_p} \times E_p$$

$$E_s = \frac{100 \text{ turns}}{20 \text{ turns}} \times 10 \text{ V}$$

$$E_s = 5 \times 10 \text{ V} = 50 \text{ V}$$

[If you are studying for an Element 3B (General) exam, study question 3E-15.1 in Chapter 11.]

X = REACTANCE

A coil or a capacitor will respond to an alternating current by storing and releasing electrical or magnetic energy as the current passes through the component. As the amount of stored energy changes, there is an opposition to the flow of more current. This opposition is called reactance, and it is measured in ohms. In formulas, reactance is designated by the letter X. Capacitive reactance is designated X_C and inductive reactance is designated X_L. Unlike resistance, the reactance of a component will vary with the frequency of the applied alternating current.

Capacitive Reactance

A discharged or "empty" capacitor looks like a short circuit. This means that when the circuit is first energized, a lot of current will flow into the capacitor as it charges. Initially, the voltage across the capacitor is zero; as the voltage builds up, the current through the capacitor will drop toward zero. See Figure 5-20. When dc is applied to a capacitor there is an initial rush of current and then the current drops to zero. The capacitor blocks the dc.

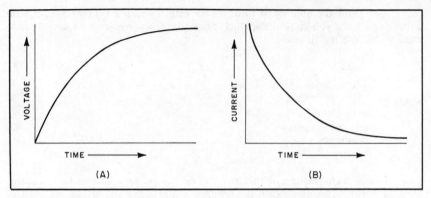

Figure 5-20—When a circuit containing a capacitor is first energized, the voltage across the capacitor is zero, and the current is very large. As time passes, the voltage across the capacitor increases, as shown at A, and the current drops toward zero, as shown at B.

When an alternating current is applied to a capacitor, things are a little different. If the frequency of the ac is low, the capacitor has time to charge to the full applied voltage during each half cycle. Once the capacitor is charged, it opposes any further increase in voltage. On the next half cycle the capacitor discharges and is then recharged with the opposite polarity. A high-frequency ac wave will not fully charge the capacitor during each half cycle, and the capacitor will pass the full current. We can see from this that the opposition to current (the reactance) decreases as the frequency of the applied wave increases.

The reactance of a capacitor is given by the formula:

$$X_C = \frac{1}{2\pi f C}$$
(Equation 5-20)

where

X_C = capacitive reactance in ohms
f   = frequency in hertz
C   = capacitance in farads
π = 3.14

You should always use the proper units. If capacitance is given in microfarads, you will have to convert to farads before using Equation 5-20. (To convert microfarads to farads, multiply microfarads by 10^{-6}. This means moving the decimal point 6 places to the left.) The equation shows the inverse relationship between capacitive reactance and frequency; as the frequency goes up, the reactance will go down.

Inductive Reactance

An inductor presents a large opposition to a change in current. This means that when a circuit containing an inductor is first energized, no current will flow in the inductor. All the applied voltage appears across the inductor; as the current increases, the voltage drops toward zero. See Figure 5-21.

Inductors pass dc. When an alternating current is applied to an inductor, the opposition to the current (the reactance) increases as the frequency increases. Inductors oppose a change in current. The higher the frequency of the applied ac, the more rapidly the current is changing, and the greater the opposition to the current. The formula for inductive reactance is:

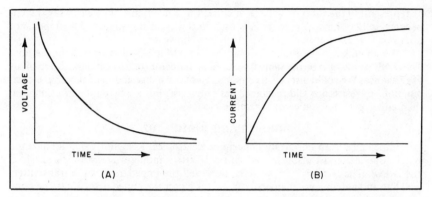

Figure 5-21—When a circuit containing an inductor is first energized, the initial current is zero and the full applied voltage appears across the inductor. As time passes, the voltage drops toward zero as shown at A, and the current increases, as shown at B.

$$X_L = 2\pi fL \hspace{4cm} \text{(Equation 5-21)}$$

where
$\hspace{1cm} X_L$ = inductive reactance in ohms
$\hspace{1cm} f \hspace{0.5cm}$ = frequency in hertz
$\hspace{1cm} L \hspace{0.5cm}$ = inductance in henrys
$\hspace{1cm} \pi \hspace{0.5cm}$ = 3.14

Again, the correct units must be used. If the inductance is expressed in units other than henrys (such as millihenrys) you will have to change them to the basic unit. From the formula, we can see that inductive reactance increases as the frequency of the applied ac increases.

[If you are studying for an Element 3B (General) exam, turn to Chapter 11 and study the questions in subelement 3E that begin 3E-3. Review this section if you have any difficulty.]

Z = IMPEDANCE

Impedance is a number obtained by dividing the voltage applied to a circuit by the current flowing through it. See Figure 5-22. Suppose we measure the current, I, flowing into the black box and find it to be 2 amperes with 250-V ac applied. Dividing 250 volts by 2 amps results in an impedance of 125. Even though this is not a dc circuit, we say that we have 125 "ohms" since the ohm has already been established

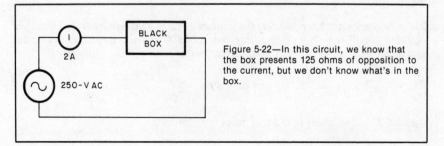

Figure 5-22—In this circuit, we know that the box presents 125 ohms of opposition to the current, but we don't know what's in the box.

as representing the ratio of voltage to current in a dc circuit. So we can say the impedance is 125 ohms. The letter Z is used to represent impedance. The impedance does not tell us what is inside the black box, however.

We could get this result (125 ohms) if we had a 125-ohm resistor in the box. The result would also be the same if we used an inductor (or capacitor) with 125 ohms of reactance. We could also use a combination of resistance and reactance. Such combinations of resistance and reactance not only exist, but are actually more common than either pure resistance or reactance alone.

Impedance Matching

Impedance is an important concept in electronics. Remember that an antenna system operates most efficiently when the SWR (standing-wave ratio) on the feed line is low. The SWR is a measure of how well the impedance of the transmitter, feed line and antenna are matched; that is, the closer the impedances are to the same value, the better the antenna system will operate.

Maximum power transfer occurs when the impedance of the source is the same as the impedance of the load. Some devices (like antennas) require a specific input impedance for proper operation; this input impedance may not be the same as the optimum output impedance for the power source (transmitter). The impedance of the source must be matched to the impedance of the load. One way to do this is with a transformer. This is called *impedance matching*.

In an ideal transformer—one without losses—the following relationship is true:

$$Z_p = Z_s \left(\frac{n_p}{n_s} \right)^2 \qquad \text{(Equation 5-22)}$$

where

Z_p = impedance seen by a device connected to the primary
Z_s = impedance seen by a device connected to the secondary
n_p = number of turns on the primary
n_s = number of turns on the secondary

From this equation, we can derive an equation for the turns ratio necessary to match a given input and output impedance.

$$\frac{n_p}{n_s} = \sqrt{\frac{Z_p}{Z_s}} \qquad \text{(Equation 5-23)}$$

In other words, the turns ratio of the transformer must equal the square root of the impedance ratio. So, if we have a 30-ohm output to match to a 30-kilohm input, the ratio of n_p to n_s is:

$$\frac{n_p}{n_s} = \sqrt{\frac{Z_p}{Z_s}} = \sqrt{\frac{30,000}{30}} = \sqrt{1000} = 31.62$$

So the primary must have 32 times the number of turns as the secondary. To match an audio amplifier with an output impedance of 600 ohms to a speaker having an impedance of 8 ohms, we will need a transformer with a turns ratio of:

$$\frac{n_p}{n_s} = \sqrt{\frac{Z_p}{Z_s}} = \sqrt{\frac{600}{8}} = \sqrt{75} = 8.67$$

Figure 5-23 shows another example using this formula.

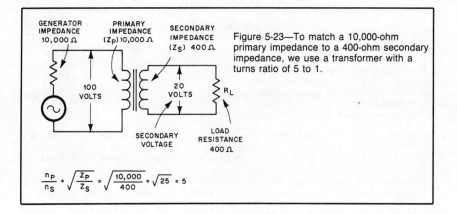

Figure 5-23—To match a 10,000-ohm primary impedance to a 400-ohm secondary impedance, we use a transformer with a turns ratio of 5 to 1.

$$\frac{n_P}{n_S} = \sqrt{\frac{Z_P}{Z_S}} = \sqrt{\frac{10,000}{400}} = \sqrt{25} = 5$$

[If you are studying for an Element 3A (Technician) exam, turn to Chapter 10 and study questions 3E-7.1 and 3E-7.3

If you are studying for an Element 3B (General) exam, study the following questions in Chapter 11: 3E-1.1, 3E-1.2, questions 3E-6.1 through 3E-6.4, 3E-7.2, 3E-7.4 and questions 3E-15.2 through 3E-15.4.]

Key Words

Avalanche point—the reverse bias voltage at which diode leakage current increases rapidly

Breakdown voltage—the voltage at which an insulating material will conduct current

Carbon-composition resistor—an electronic component designed to limit current in a circuit; made from ground carbon mixed with clay

Capacitor—an electronic component composed of two or more conductive plates separated by an insulating material

Ceramic capacitor—a capacitor formed by depositing metal electrodes on both sides of a ceramic disk

Coefficient of coupling—a measure of the mutual inductance between two coils. The greater the coefficient of coupling, the more mutual inductance between the coils.

Color code—colored stripes painted on the body of a resistor to show the value of the resistor

Core—the material used in the center of an inductor coil

Dielectric—the insulating material used between the plates in a capacitor

Dielectric constant—a number used to indicate the relative "merit" of an insulating material. Air is given a value of 1, and all other materials are related to air.

Distributed capacitance—the capacitance that exists between turns of a coil

Electrolytic capacitor—a polarized capacitor formed using thin foil electrodes and chemical-soaked paper

Electric field—an invisible force of nature. An electric field exists in a region of space if an electrically charged object placed in the region is subjected to an electrical force

Forward biased—a condition in which the anode of a diode is positive with respect to the cathode, so it will conduct

Heat sink—a piece of metal used to absorb excess heat generated by a component, and dissipate the heat into the air

Junction diode—an electronic component formed by placing a layer of N-type semiconductor material next to a layer of P-type material. Diodes allow current to flow in one direction only.

Maximum average forward current—the highest average forward current that can flow through a diode for a given junction temperature

Metal-film resistor—a resistor formed by depositing a thin layer of resistive-metal alloy on a cylindrical ceramic form

Mica capacitor—a capacitor formed by alternating layers of metal foil with thin sheets of insulating mica

Mutual inductance—when coils display mutual inductance, a current flowing in one coil will induce a voltage in the other. The magnetic flux of one coil passes through the windings of the other.

N-type material—semiconductor material that has been treated with impurities to give it an excess of electrons

P-type material—semiconductor material that has been treated with impurities to give it an electron shortage

Peak inverse voltage (PIV)—the maximum voltage a diode can withstand when it is reverse biased (not conducting)

Plastic-film capacitor—a capacitor formed by sandwiching thin sheets of mylar or polystyrene between thin foil plates, and rolling the entire unit into a cylinder

Potentiometer—a resistor whose resistance can be varied continuously over a range of values

Primary winding—the coil in a transformer that is connected to the energy source

Resistor—any material that opposes a current in an electrical circuit. An electronic component specifically designed to oppose current

Reverse biased—a condition in which the cathode of a diode is positive with respect to the anode, so it will not conduct

Rotor—the movable plates in a variable capacitor

Secondary winding—the coil in a transformer that is connected to the load

Semiconductor—a material whose conductive properties can be controlled by adding impurities so it can act as a good conductor sometimes and as a good insulator at other times

Stator—the stationary plates in a variable capacitor

Temperature coefficient—a number used to show whether a component will increase or decrease in value as it gets warm

Toroid—a coil wound on a donut-shaped ferrite or powdered-iron form

Transformer—mutually coupled coils used to change the voltage level of an ac power source to one more suitable for a particular circuit

Variable capacitor—a capacitor that can have its value changed within a certain range

Wire-wound resistor—a resistor made by winding a length of wire on an insulating form

Variable resistor—a resistor whose value can be adjusted over a certain range

Chapter 6

Circuit Components

B efore you can understand the operation of most complex electronic circuits, you must know some basic information about the parts that make up those circuits. This chapter presents the information about circuit components that you need to know to pass your Technician/General class written exam. You will find descriptions of resistors, capacitors, inductors and transformers, along with some basic information about semiconductor diodes. These components are combined with other devices to build practical electronic circuits, some of which are described in Chapter 7.

As you study the characteristics of the components described in this chapter, be sure to turn to the FCC Element 3 questions in Chapter 10 when you are directed to do so. That will show you where you need to do some extra studying. If you thoroughly understand how these components work, you should have no problem learning how they can be connected to make a circuit perform a specific task.

RESISTORS

We have seen that current in a circuit is the flow of electrons from one point to another. A perfect insulator would allow no electron flow (zero current), while a perfect conductor would allow infinite electron flow (infinite current). In practice, however, there is no such thing as a perfect conductor or a perfect insulator. Partial opposition to electron flow occurs when the electrons collide with other electrons or atoms in the conductor. The result is a reduction in current, and heat is produced in the conductor. *Resistors* allow us to control the current in a circuit by controlling the opposition to electron flow. As they oppose the flow of electrons they dissipate electrical energy in the form of heat. The more energy a resistor dissipates, the hotter it will become.

All conductors exhibit some resistance—Table 6-1 shows the resistance in ohms per 1000 feet of some common copper and nickel wire sizes. We can see that the smaller the conductor, the greater the resistance. This is just what we would expect if we think of

Table 6-1

Wire Resistance Per 1000 Feet

Wire Size	Diam (inches)	Material	Ohms per 1000 ft at 25° C
20	0.032	copper	10.35
22	0.025		16.46
24	0.020		26.17
26	0.016		41.62
28	0.013		66.17
30	0.010	↓	105.2
20	0.032	nickel	52.78
22	0.025		83.95
24	0.020		133.47
26	0.016		212.26
28	0.013		337.47
30	0.010	↓	536.52

electron flow in a conductor as being like water flow in a pipe. If we reduce the size of the pipe, not as much water can flow through it.

One simple way of producing a resistor is to use a length of wire. For example, we can see from Table 6-1 that the resistance of 1000 feet of number 28 nickel wire is 337 ohms. Therefore, if we need a resistor of 34 ohms, we can use 100 feet of this size nickel wire. A 100-foot length of wire is a little difficult to work with, but if we wind the wire over a form of some sort, we have a unit of a much more convenient size. This is precisely how *wire-wound resistors* are constructed. See Figure 6-1. The problem with this type of construction is that at radio frequencies, the wire-wound construction causes the resistor to act as an inductor and not as a pure resistance. For RF circuits, we must have resistors that are noninductive.

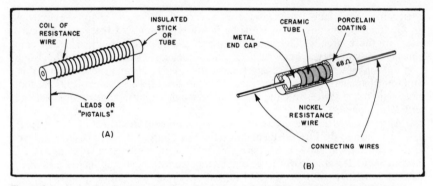

Figure 6-1—A simple wire-wound resistor is shown at A. At B is a commercially produced wire-wound resistor. These resistors are produced by winding resistive wire around a nonconductive form. Most wire-wound resistors are then covered with a protective coating.

Another way to form a resistor is to connect leads to both ends of a block or cylinder of material that has a high resistance. Figure 6-2 shows a resistor made in this manner using a mixture of carbon and clay as the resistive element. This type of resistor is called a *carbon-composition resistor*. The proportions of carbon and clay determine the value of the resistor.

Some advantages of carbon-composition resistors are the wide range of available values, low inductance and capacitance, good surge-handling capability and the ability to withstand small power overloads without being completely destroyed. The main disadvantage is that the resistance of the composition resistor will vary widely as the operating temperature changes and as the resistor ages.

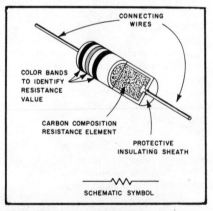

Figure 6-2—Composition resistors are composed of a mixture of carbon and clay. The resistance is controlled by varying the amount of carbon in the material when the resistor is manufactured.

The *metal-film resistor* has replaced the carbon-composition type in many low-power applications today. Metal-film resistors are formed by depositing a thin layer

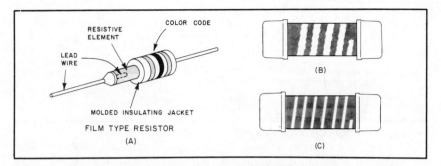

Figure 6-3—The construction of a typical metal-film resistor is shown at A. Film resistors are formed by depositing a thin film of material on a ceramic form. Material is then trimmed off in a spiral to produce the specified resistance for that particular resistor. The excess material may either be trimmed with a mechanical lathe or a laser. A lathe produces a rather rough spiral, like that shown in part B. The laser produces a finer cut, shown at C.

of Nichrome (an alloy made of nickel and chromium) or other resistive alloy on a cylindrical ceramic form. See Figure 6-3. This film is then trimmed away in a spiral fashion to form the resistance path. The trimming can be done on a mechanical lathe or by using a laser. The resistor is then covered with an insulating material to protect it.

Because of the spiral shape of the resistance path, metal-film resistors will exhibit some inductive properties, especially at high frequencies. The higher the resistance of the unit, the greater the inductance will be, due to the increased path length. Metal-film resistors provide much better temperature stability than other types. They are easily recognized by their "dogbone" shape—they are thinner in the middle than at the ends, because of the metal end caps used to connect the leads to the resistive element.

Most resistors have standard fixed values, so they can be called fixed resistors. *Variable resistors* (sometimes called *potentiometers* because they can be used to adjust the voltage, or potential, in a circuit) are also used a great deal in

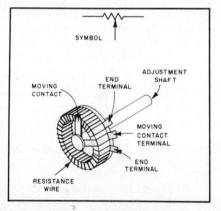

Figure 6-4—The typical construction of a wire-wound variable resistor.

electronics. The construction of a wire-wound variable resistor and its schematic symbol are shown in Figure 6-4. Variable resistors may also be made with a ring of carbon compound in place of the wire windings. A connection is made to each end of the resistance ring, and a third contact is attached to a movable arm, or wiper, that can be moved across the ring. As the wiper moves from one end of the ring to the other, the resistance varies from minimum to maximum.

Color Codes

Standard fixed resistors can usually be found in values ranging from 2.7 Ω to 22 MΩ (22 megohms, or 22,000,000 ohms). Resistance tolerances on these standard values can be ±20%, ±10%, ±5% and ±1%. A tolerance of 10% on a 200-ohm resistor means that the actual resistance of a particular unit may be anywhere from 180 Ω to 220 Ω. (Ten percent of 200 is 20, so the resistance can be 200 + 20 or 200 − 20 ohms.)

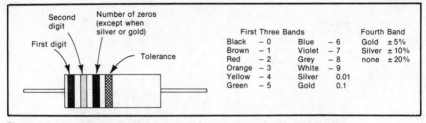

Figure 6-5—Small resistors are labeled with a color code to show their value.

Because it is not practical to print the resistance value on the side of a small resistor, a *color code* is used to show the value of the resistor. As shown in Figure 6-5, the color of the first three bands shows the resistor value, and the color of the fourth band indicates the resistor tolerance.

Power Ratings

Resistors are also rated according to the amount of power they can safely handle. Power ratings for wire-wound resistors typically start at 1 watt and range to 10 W or larger. Resistors with higher power ratings are also physically large—greater surface area is required to dissipate the increased heat generated by large currents. Carbon-composition and film resistors are low-power components. You will find these resistors in 1/8-watt, 1/4-W, 1/2-W, 1-W and 2-W power ratings. A resistor that is being pushed to the limit of its power rating will feel very hot to the touch. If a resistor is getting very hot during normal operation, it should probably be replaced with a unit having a higher power-dissipation rating.

Temperature Coefficient

As the temperature of a resistor changes, its resistance will change. How the resistance will change is determined by the *temperature coefficient* of the resistor. With a positive temperature coefficient, the resistance will increase as the temperature increases. With a negative coefficient, the resistance will decrease as the temperature increases. Some types of resistors have positive coefficients, and some types have negative coefficients. One thing is certain—as the temperature changes, the value of the resistor will change.

[If you are studying for an Element 3A (Technician) exam, turn to Chapter 10 and study questions 3F-1.1 through 3F-1.4 and question 3F-1.6.

If you are studying for an Element 3B (General) exam, study question 3F-1.5 in Chapter 11.]

CAPACITORS

In Chapter 5 we learned that the basic property of a *capacitor* is the ability to store an electric charge. In an uncharged capacitor, the potential difference between the two plates is zero. As we charge the capacitor, this potential difference increases until it reaches the full applied voltage. The difference in potential creates an *electric field* between the two plates. See Figure 6-6. The electric field between the plates of our capacitor is invisible—a field is an invisible force of nature. We put energy into the capacitor by charging it, and until we discharge it or the charge leaks away somehow, the energy is stored in the electric field. When the field is not moving we sometimes call it an electrostatic field.

When a dc voltage is applied to a capacitor, current will flow in the circuit until the capacitor is fully charged. After the capacitor has charged to the full dc voltage, no more current will flow in the circuit. For this reason, capacitors can be used to

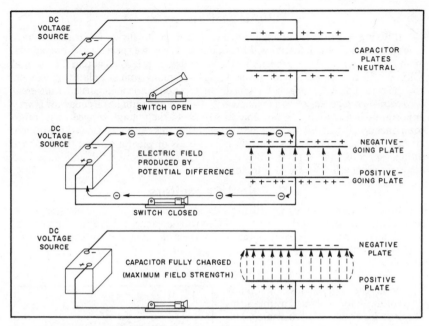

Figure 6-6—Capacitors store energy in an electric field.

block dc in a circuit. Because ac voltages are constantly reversing polarity, we can block dc with a capacitor while permitting ac to pass.

Factors That Determine Capacitance

Remember from Chapter 4 that the value of a capacitor is determined by three factors: the area of the plate surfaces, the distance between the plates and the type of insulating material used between the plates. As the surface area is increased, the capacitance increases. For a given plate area, as the spacing between the plates is decreased, the capacitance increases.

For a given value of plate spacing and area, the capacitance is determined by the type of dielectric material used in the capacitor. A capacitor using air as the dielectric is used for comparison purposes, and other dielectric materials are related to the standard air-dielectric capacitor by the *dielectric constant* of the material used. See Table 6-2. From the chart we can see that if polystyrene is used as the dielectric in a capacitor, it will have 2.6 times the capacitance of an air-dielectric capacitor with the same plate spacing and area.

Table 6-2

Dielectric Constants and Breakdown Voltages

Material	Dielectric Constant*	Breakdown Voltage**
Air	1.0	21
Alsimag 196	5.7	240
Bakelite	4.4-5.4	300
Bakelite, mica filled	4.7	325-375
Cellulose acetate	3.3-3.9	250-600
Fiber	5-7.5	150-180
Formica	4.6-4.9	450
Glass, window	7.6-8	200-250
Glass, Pyrex	4.8	335
Mica, ruby	5.4	3800-5600
Mycalex	7.4	250
Paper, Royalgrey	3.0	200
Plexiglas	2.8	990
Polyethylene	2.3	1200
Polystyrene	2.6	500-700
Porcelain	5.1-5.9	40-100
Quartz, fused	3.8	1000
Steatite, low loss	5.8	150-315
Teflon	2.1	1000-2000

*At 1 MHz   **In volts per mil (0.001 inch)

Voltage Ratings

If we increase the voltage applied to a capacitor, eventually we will reach a voltage at which the dielectric material will break down. When we speak of the breakdown of an insulating material, we mean that the applied voltage is so great that the material will conduct. In a capacitor, this means a spark will jump from one plate to the other, and the capacitor will probably be ruined. Different dielectric materials have different *breakdown voltages*, as shown in Table 6-2. The voltage rating of a capacitor is very important—if a capacitor is not able to withstand the voltage applied to it, other components in the circuit may also be damaged. For this reason, capacitors are usually labeled with both their capacitance in microfarads or picofarads and their voltage rating, specified in working-volts dc (WVDC). Sometimes the tolerance and temperature coefficient are also indicated.

Practical Capacitors

Practical capacitors are described by the material used for their dielectric. Mica, ceramic, mylar, polystyrene, paper and electrolytic capacitors are in common use today. They each have properties that make them more or less suitable for a particular application.

Mica Capacitors

Mica capacitors consist of many strips of metal foil separated by thin strips of mica. See Figure 6-7. Alternate plates are connected together and each set of plates is connected to an electrode. The entire unit is then encased in plastic or ceramic insulating material. An alternative to this form of construction is the "silvered-mica" capacitor. In the silvered-mica type, a thin layer of silver is deposited directly onto one side of the mica, and the plates are stacked so that alternate layers of mica are separated by layers of silver.

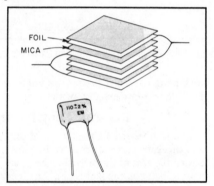

Figure 6-7—Mica capacitors are formed by interleaving metal foil with thin sheets of mica.

From Table 6-2 we can see that mica has a very high voltage breakdown rating. For this reason, mica capacitors are frequently used in transmitters and high-power amplifiers, where the ability to withstand high voltages is important. Mica capacitors also have good temperature stability—their capacitance does not change greatly as the temperature changes. Typical capacitance values for mica capacitors range from 1 picofarad to 0.1 microfarad, and voltage ratings as high as 35,000 are possible.

Ceramic Capacitors

Ceramic capacitors are constructed by depositing a thin metal film on each side of a ceramic disc, as shown in Figure 6-8. Wire leads are then attached to the metal films, and the entire unit is covered with a protective plastic or ceramic coating. Ceramic capacitors are inexpensive and easy to construct, and they are in wide use today.

Ordinary ceramic capacitors cannot be used where temperature stability is important—their capacitance will change with a change in temperature. Special ceramic capacitors (called NP0 for negative-positive zero) are used for these applications. The capacitance of an NPØ unit will remain substantially the same over a wide range of temperatures. The range of capacitance values available with ceramic capacitors is typically 1 picofarad to 0.1 microfarad, with working voltages up to 1000.

Ceramic capacitors are often connected across the transformer primary or secondary winding in a power supply. These capacitors, referred to as suppressor capacitors, are intended to suppress any transient voltage spikes, preventing them from getting through the power supply.

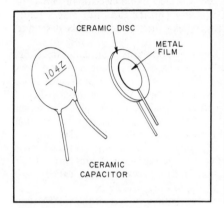

Figure 6-8—In the ceramic capacitor, electrodes are deposited on both sides of a ceramic disc.

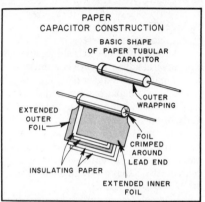

Figure 6-9—Paper tubular capacitors are formed of rolled foil with insulating paper layers.

Paper Capacitors

In its simplest form, the paper capacitor consists of a layer of paper between two layers of metal foil. The foil and paper are rolled up, as shown in Figure 6-9. Wires are connected to the foil layers, and the capacitor is encased in plastic or dipped in wax to protect it. One end of the capacitor sometimes has a band around it. This does not mean that the capacitor is polarized, only that the outside layer of foil is connected to the marked lead. This way, the outside of the capacitor can be grounded where necessary for RF shielding.

Paper capacitors can be obtained in capacitance values from about 500 picofarads to around 50 microfarads, and in voltage ratings up to about 600 WVDC. Paper capacitors are generally inexpensive, but their large size for a given value makes them impractical for some uses.

Plastic-Film Capacitors

Construction techniques similar to those used for paper capacitors are used for *plastic-film capacitors*. Thin sheets of mylar or polystyrene are sandwiched between sheets of metal foil, and the construction is rolled into a cylinder. The plastic material gives capacitors with a high voltage rating in a physically small package. Plastic-film capacitors also have good temperature stability; typical values range from 5 picofarads to 0.47 microfarad.

Electrolytic Capacitors

In *electrolytic capacitors*, the dielectric is formed after the capacitor is manufactured. The construction of aluminum-electrolytic capacitors is very similar to that of paper capacitors. Two sheets of aluminum foil separated by paper soaked in a chemical solution are rolled up and placed in a protective casing. After assembly, a voltage is applied to the capacitor causing a thin layer of aluminum oxide to form on the surface of the positive plate next to the chemical. The aluminum oxide acts as the dielectric, and the foil plates act as the electrodes.

Because of the extremely thin layer of oxide dielectric, electrolytic capacitors can be made with a very high capacitance value in a small package. Electrolytic capacitors are polarized—dc voltages must be connected to the positive and negative capacitor terminals with the correct polarity. The positive and negative electrodes in an electrolytic capacitor are clearly marked. Connecting an electrolytic capacitor incorrectly causes gas to form inside the capacitor and the capacitor may actually explode. This can be very dangerous. At the very least the capacitor will be destroyed by connecting it incorrectly.

Another type of electrolytic capacitor is gaining popularity. Tantalum capacitors have several advantages over aluminum-electrolytic capacitors. Tantalum capacitors can be made even smaller than aluminum-electrolytic capacitors for a given capacitance value. They are manufactured in several forms, including small, water-droplet-shaped solid-electrolyte capacitors. These are formed on a small tantalum pellet that serves as the anode, or positive capacitor plate. An oxide layer on the outside of the tantalum pellet serves as the dielectric, and a layer of manganese dioxide is the solid electrolyte. Layers of carbon and silver form the cathode, or negative capacitor plate. The entire unit is dipped in epoxy to form a protective coating on the capacitor. Their characteristic shape explains why these tantalum capacitors are often called "tear drop" capacitors.

Electrolytic capacitors are available in voltage ratings of greater than 400 V, and capacitance values from 1 microfarad to 100,000 microfarads (0.1 farad). Electrolytic capacitors with high capacitance values and/or high voltage ratings are physically very large. Electrolytic capacitors are used in power-supply filters, where large values of capacitance are necessary for good smoothing of the pulsating dc from the rectifier.

Variable Capacitors

In some circuits it is necessary or desirable to be able to vary the capacitance at some point; for example, in the tuning circuit of a receiver or transmitter VFO. We could use a rotary switch to select one of several different fixed-value capacitors, but it is much more convenient to use a *variable capacitor*. A basic air-dielectric variable capacitor is shown in Figure 6-10. One set of plates (called the *stator*) is fixed

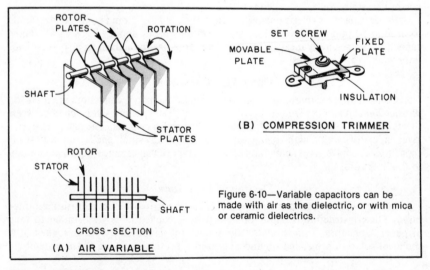

Figure 6-10—Variable capacitors can be made with air as the dielectric, or with mica or ceramic dielectrics.

and the other set of plates (called the *rotor*) can be rotated to control the amount of plate area shared by the two sets of plates. When the capacitor is fully meshed (all the rotor plates are down in between the stator plates) the capacitor will have its largest value of capacitance. When the rotor is unmeshed, the value of capacitance is at a minimum. The capacitance can be varied smoothly between the maximum and minimum values.

Another type of variable capacitor is the compression variable. In this type of variable capacitor, the spacing between the two plates is varied. When the spacing is at its minimum, the capacitance is at a maximum value. When the spacing is adjusted to its maximum, the value of capacitance is at a minimum. This type of variable capacitor is usually used as what is known as a *trimmer capacitor*—a control used to peak or fine tune a part of a circuit, and left alone once adjusted to the correct value.

Variable capacitors are subject to the same rules as other capacitors. If a variable capacitor is to be used with a high voltage on its plates, a large plate spacing or a dielectric with a high breakdown voltage will be required. Air-variable capacitors used in transmitters and high-power amplifiers have their plates spaced much farther apart than those used in receivers.

[If you are studying for an Element 3A (Technician) exam, turn to Chapter 10 and study questions 3F-2.1 through 3F-2.5 and question 3F-2.8.

If you are studying for an Element 3B (General) exam, study questions 3F-2.6 and 3F-2.7 in Chapter 11.]

INDUCTORS

Just as a capacitor stores energy in an electric field, an inductor stores energy in a magnetic field. When a voltage is first applied to an inductor, with no current flowing through the circuit, the inductor will oppose the current. Remember from Chapter 5 that inductors oppose any change in current. A voltage is induced in the coil that opposes the applied voltage, and tries to prevent a current. Gradually the current will build up to a value that is limited only by any resistance in the circuit (the resistance in the wire of the coil will be very small). In the process of getting this current to flow, energy is stored in the coil in the form of a magnetic field around the wire. When the applied current is shut off, the magnetic field collapses and returns the stored energy to the circuit as a current flowing in the opposite direction. This reverse current is produced by the back EMF generated by the collapsing magnetic field. Again, this voltage opposes any change in the current already flowing in the coil. The amount of opposition to changes in current, called reactance, the amount of energy stored in the magnetic field and the back EMF induced in the coil all depend on the inductance of the coil.

The amount of inductance that a coil exhibits, represented by the symbol L, depends on four things:

1) the type of material used in the core, and its size and location in the coil
2) the number of turns used to wind the coil
3) the length of the coil and
4) the diameter of the coil.

Changing any of these factors changes the inductance.

If an iron or ferrite core is added to the coil, the inductance increases. Brass can also be used as a core material in radio-frequency variable inductors. It has the opposite effect of iron or ferrite: The inductance decreases as the brass core enters the coil. If a movable core is placed in a coil, as shown in Figure 6-11, the amount of inductance can be varied as the core is moved in and out of the coil. The common schematic symbols for various inductors are shown in Figure 6-12.

Ferrite comes from the Latin word for iron. So what is the difference between an iron and a ferrite core? Well, iron cores can be made from iron sheet metal, such as is used in the layers of material that make up a transformer coil, or they can be

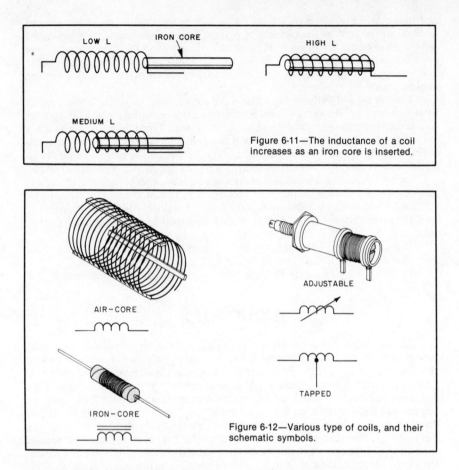

Figure 6-11—The inductance of a coil increases as an iron core is inserted.

Figure 6-12—Various type of coils, and their schematic symbols.

made from powdered iron that has been mixed with a bonding material to hold it together and then molded into the desired shape. If other metal alloys are mixed with the iron "ferrous" material, then the core is referred to as a ferrite core. By selecting the proper alloy to provide the desired characteristics, the core can be designed to operate best over a specific frequency range. Iron and powdered-iron cores are used most often at low frequencies, such as in the audio-frequency range and ac-power circuits. Ferrite cores are manufactured in a wide range of compositions, to optimize their operation for specific radio-frequency ranges.

It takes some amount of energy to magnetize the core material for any inductor, and this energy represents a loss in the coil. When the coil is wound on an iron material that increases the inductance of the coil, the energy needed to magnetize the core also increases. Coils wound on a thin plastic form, or made self supporting so the only material inside the coil is air, are called air-wound coils. These have the lowest loss of any type of inductor, and are normally used in high-power RF circuits where energy loss must be kept to a minimum.

Toroid Cores

A *toroid* is a coil wound on a core that is curved into a donut shape. Powdered-iron or ferrite compounds are usually used for toroidal core materials. The toroid is a highly efficient inductor, because there is no break in the circular core. All of the magnetic lines of force remain inside the core. This means there is very little mutual coupling between two toroids mounted close to each other in a circuit. See Figure 6-13. Mutual coupling can be a problem, especially in an RF circuit where signals from one stage could be coupled into another stage without following the proper signal path. Each stage can be built into its own shielded compartment, but using toroid cores can often help eliminate the mutual coupling that would occur between other types of coils.

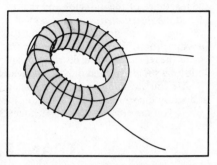

Figure 6-13—A toroidal coil. This type of coil has self-shielding properties.

In theory, a toroid requires no shield to prevent its magnetic field from spreading out and interfering with outside circuits. This self-shielding property also helps keep outside forces from interfering with the toroidal coil. Toroids can be mounted so close to each other that they can almost touch, but because of the way they are made, there will be almost no inductive coupling between them. With ferrite or iron cores, toroidal coils wound with only a small amount of wire can have a very high inductance value.

Self-resonance

Any two conductors placed near each other with an insulating material between them will form a capacitor. In the case of a coil, the turns of the coil act as if they had tiny capacitors connected between them. See Figure 6-14. At some frequency,

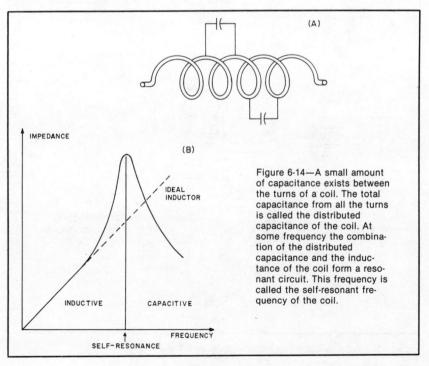

Figure 6-14—A small amount of capacitance exists between the turns of a coil. The total capacitance from all the turns is called the distributed capacitance of the coil. At some frequency the combination of the distributed capacitance and the inductance of the coil form a resonant circuit. This frequency is called the self-resonant frequency of the coil.

the reactance of the inductor and the reactance of the capacitors between the turns (called the *distributed capacitance*) will be equal. In an LC circuit at resonance, the reactance of the inductor (or inductors) equals the reactance of the capacitor (or capacitors). With an inductor, the frequency where the reactance of the inductor equals the reactance of the distributed capacitance is called the self-resonant frequency. Inductors should be used only at frequencies well below their self-resonant frequency.

[If you are studying for an Element 3A (Technician) exam, turn to Chapter 10 and study questions 3F-3.1 through 3F-3.6.

If you are studying for an Element 3B (General) exam, study question 3F-3.5 in Chapter 11.]

TRANSFORMERS

Remember that when coils have mutual inductance the flux lines from one coil cut the turns of the other coil, and a changing current in one coil can induce a voltage in the other. Therefore, two coils having mutual inductance can form a *transformer*. The coil connected to the source of energy is called the *primary winding*, and the coil connected to the load is called the *secondary winding*.

With a transformer, electrical energy can be transferred from one circuit to another without direct connection. In the process, the signal voltage can be readily changed from one level to another. Common house current is 117-volts ac. If you

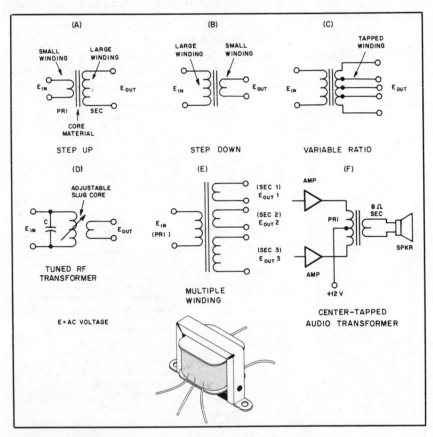

Figure 6-15—Schematic symbols for various types of transformers.

need 234-volts ac, you would use a transformer that has twice as many turns on the secondary coil as it does on its primary coil. The induced voltage in the secondary would then be double the primary voltage, or 234 volts. A transformer can only be used with alternating current, because no voltages will be induced in the secondary if the magnetic field is not changing.

The primary and secondary coils of a transformer may be wound on a core of a magnetic material, such as iron. This increases the coupling of the coils so that practically all of the field set up by the current in the primary coil will cut the turns of the secondary coil. A small amount of current, called the magnetizing current, flows in a transformer primary winding, even with no load connected to the secondary. This magnetizing current is necessary to overcome losses in the iron core and the resistance of the primary winding, and represents energy lost in the transformer. Schematic symbols for some different types of transformers are shown in Figure 6-15.

Transformer Ratings

Transformers have power ratings, just as other components do. Usually, the primary and secondary voltages are specified, along with the maximum current that can be drawn from the secondary winding. Sometimes the secondary voltage and current are combined into a "volt-ampere" (VA) rating. A transformer with a 12-V, 1.5-A secondary would have a volt-ampere rating of 18 VA.

[If you are studying for an Element 3B (General) exam, turn to Chapter 11 and study the questions in subelement 3F that begin 3F-4.]

DIODES

Semiconductor material exhibits properties of both metallic and nonmetallic substances. Generally, the layers of semiconductor material are made from germanium or silicon. To control the conductive properties of the material, impurities are added as the semiconductor crystal is "grown." This process is called doping. If a material with an excess of free electrons is added to the structure, *N-type material* is produced. This type of material is sometimes referred to as donor material. When an impurity that results in an absence of electrons is introduced into the crystal structure, you get *P-type material*. The result in this case is a material with positive charge carriers, called holes. P-type material is also called acceptor material.

The *junction diode*, also called the PN-junction diode, is made from two layers of semiconductor material joined together. One layer is made from P-type material and the other layer is made from N-type material. The name PN junction comes from the way the P and N layers are joined to form a semiconductor diode. Figure 6-16 illustrates the basic concept of a junction diode.

The P-type side of the diode is called the *anode*, while the N-type side is called the *cathode*. (The anode connects to the positive supply lead and the cathode connects to the negative supply lead.) Current flows from the cathode to the anode; that is, the excess electrons from the N-type material flow to the P-type material, which has holes (an electron deficiency). Electrons and holes are called charge carriers because they carry current from one side of the junction to the other. When no voltage is

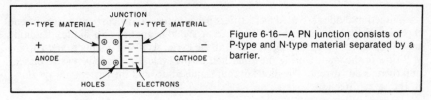

Figure 6-16—A PN junction consists of P-type and N-type material separated by a barrier.

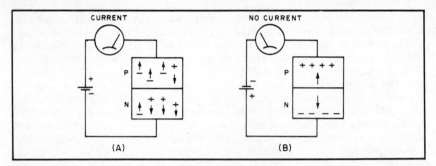

Figure 6-17—At A, the PN junction is forward biased and is conducting. At B, the junction is reverse biased and it does not conduct.

applied to a diode, the junction between the N-type and P-type material acts as a barrier that prevents carriers from flowing between the layers.

When voltage is applied to a junction diode as shown at A in Figure 6-17, however, carriers will flow across the barrier and the diode will conduct (electrons will flow through it). When the diode anode is positive with respect to the cathode, electrons are attracted across the barrier from the N-type material, through the P-type material and on to the positive battery terminal. As the electrons move through the P-type material, they fill some holes and leave other holes behind as they move, so the holes appear to move in the opposite direction, toward the negative battery terminal. When the diode is connected in this manner it is said to be *forward biased*.

If the battery polarity is reversed, as shown in Figure 6-17B, the excess electrons in the N-type material are attracted away from the junction by the positive battery terminal. Electrons in the P-type material are forced toward the junction by the negative charge from the battery, and the electrons leave holes that appear to move toward the negative battery terminal. Because there are no holes in the N-type material to accept electrons, no conduction takes place across the junction. When the anode is connected to a negative voltage source and the cathode is connected to a positive source, the diode does not conduct, and it is said to be *reverse biased*.

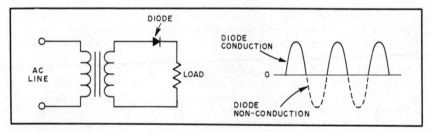

Figure 6-18—PN junction diodes are used as rectifiers because they allow current to flow in one direction only.

Junction diodes are used as rectifiers because they allow current to flow in one direction only. See Figure 6-18. When an ac signal is applied to a diode, the diode will be forward biased during one half of the cycle, so it will conduct, and current will flow to the load. During the other half of the cycle, the diode is reverse biased, and there is no current. The diode output is pulsed dc, and the current always flows in the same direction.

Diode Ratings

Junction diodes have maximum voltage and current ratings that must be observed, or damage to the junction will result. The voltage rating is called the *peak inverse voltage (PIV)* and the current rating is called the forward current. With present technology, diodes are commonly available with ratings up to 1000 PIV and 100 A forward current.

Peak inverse voltage is the voltage a diode must withstand when it isn't conducting. Although a diode is normally used in the forward direction, it will conduct in the reverse direction if enough voltage is applied. A few hole/electron pairs are thermally generated at the junction when a diode is reverse biased. These pairs cause a very small reverse current, called leakage current, to flow. Semiconductor diodes can withstand some leakage current, but if the inverse voltage reaches a high enough value the leakage current will rise abruptly, resulting in a heavy reverse current. The point at which the leakage current rises abruptly is called the *avalanche point*. A large reverse current usually damages or destroys the diode.

The maximum average forward current is the highest average current that can flow through the diode in the forward direction for a specified junction temperature. This specification varies from device to device, and it depends on the maximum allowable junction temperature and on the amount of heat the device can dissipate. As the forward current increases, the junction temperature will increase. If the diode is allowed to get too hot, it will be destroyed.

Impurities at the PN junction cause some resistance to current. This resistance results in a voltage drop across the junction. For silicon diodes, this drop is approximately 0.6 to 0.7 V; for germanium diodes, the drop is 0.2 to 0.3 V. When current flows through the junction, some power is dissipated in the form of heat. The amount of power dissipated depends on the amount of current flowing through the diode. For example, it would be approximately 6 W for a silicon rectifier with 10 A flowing through it:

$$P = I \times E$$
$$P = 10 \text{ A} \times 0.6 \text{ V} = 6 \text{ W}$$

If the junction temperature exceeds the safe level specified by the manufacturer, the diode is likely to be damaged or destroyed.

Diodes designed to safely handle forward currents in excess of 6 A generally are packaged so they may be mounted on a *heat sink*. These diodes are often referred to as stud-mount devices, because they are made with a threaded screw attached to the body of the device. This screw is attached to the heat sink so that the heat from the device will be effectively transferred to the heat sink. The heat sink helps the diode package dissipate heat more rapidly, thereby keeping the junction temperature at a safe level. The metal case of a stud-mount diode is usually one of the contact points, so it must be insulated from ground.

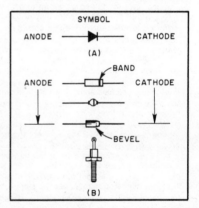

Figure 6-19—The schematic symbol for a diode is shown at A. Diodes are typically packaged in one of the case styles shown at B.

Figure 6-19 shows some of the more common diode-case styles, and the general schematic symbol for a diode. The line or spot on a diode case indicates the cathode lead. On a high-power stud-mount diode, the stud may either be the anode or the cathode. Check the case or the manufacturer's data sheet for the correct polarity.

[If you are studying for an Element 3A (Technician) exam, turn to Chapter 10 and study question 3F-5.3.

If you are studying for an Element 3B (General) exam, study questions 3F-5.1, 3F-5.2 and 3F-5.4 in Chapter 11.]

Key Words

Alternating current (ac)—electrical current that flows first in one direction in a conductor and then in the other direction. The applied voltage is also changing polarity. This direction reversal continues at a rate that depends on the frequency of the ac.

Attenuate—to reduce in amplitude

Band-pass filter—a circuit that allows signals to go through it only if they are within a certain range of frequencies, and attenuates signals above and below this range

Beat-frequency oscillator (BFO)—an oscillator that provides a signal to the product detector. In the product detector, the BFO signal and the IF signal are mixed to produce an audio signal.

Bleeder resistor—a large-value resistor connected to the filter capacitor in a power supply to discharge the filter capacitors when the supply is switched off

Capacitor-input filter—a power-supply filter having a capacitor connected directly to the rectifier output

Choke-input filter—a power-supply filter having an inductor (choke) connected directly to the rectifier output

Cutoff frequency—in a high-pass or low-pass filter, the cutoff frequency is the frequency at which the filter begins to attenuate signals

Detector—the stage in a receiver in which the modulation (voice or other information) is recovered from the RF signal

Direct-conversion receiver—A receiver that converts an RF signal directly to an audio signal with one mixing stage

Direct current (dc)—electrical current that flows in only one direction in a conductor

Filter—a circuit that will allow some signals to pass through it but will greatly reduce the strength of others

Filter choke—a large-value inductor used in a power-supply filter to smooth out the ripple in the pulsating dc output from the rectifier

Frequency multiplier—a circuit used to increase the frequency of a signal by integral multiples

Full-wave bridge rectifier—a full-wave rectifier circuit that uses four diodes and does not require a center-tapped transformer

Full-wave rectifier—a circuit basically composed of two half-wave rectifiers. The full-wave rectifier allows the full ac waveform to pass through; one half of the cycle is reversed in polarity. This circuit requires a center-tapped transformer.

Half-wave rectifier—a circuit that allows only half of the applied ac waveform to pass through it

High-pass filter—a filter that allows signals above the cutoff frequency to pass through, and attenuates signals below the cutoff frequency

Intermediate frequency (IF)—the output frequency of a mixing stage in a superheterodyne receiver; the subsequent stages in the receiver are tuned for maximum efficiency at the IF.

Low-pass filter—a filter that allows signals below the cutoff frequency to pass through and attenuates signals above the cutoff frequency

Mixer—a circuit used to combine two or more audio- or radio-frequency signals to produce a different output frequency

Oscillator—a circuit built by adding positive feedback to an amplifier. It produces an alternating-current signal with no input except the dc operating voltages.

Power supply—a device used to convert the available voltage and current source (often the 117-V ac household supply) to a form that is required for a specific circuit requirement. This will often be a higher or lower dc voltage.

Pulsating dc—the output from a rectifier before it is filtered. The polarity of a pulsating dc source does not change, but the amplitude of the voltage changes with time.

Rectifier—an electronic component that allows current to pass through it in only one direction

Ripple—the amount of change between the maximum voltage and the minimum voltage in a pulsating dc waveform

Sensitivity—the ability of a receiver to detect weak signals

Selectivity—a measure of how well a receiver can separate a desired signal from other signals on a nearby frequency

Stability—a measure of how well a receiver or transmitter will remain on frequency without drifting

Superheterodyne receiver—a receiver that converts RF signals to audio signals using at least two mixing stages

Variable-frequency oscillator (VFO)—an oscillator used in receivers and transmitters; the frequency is set by a tuned circuit using capacitors and inductors. The frequency can be changed by adjusting the components in the tuned circuit.

Chapter 7

Practical Circuits

N ow that you have studied some basic electrical principles, and have learned about the properties of some simple components, you are ready to learn how to apply those ideas to practical Amateur Radio circuits. This chapter will lead you through examples and explanations to help you gain that knowledge.

In this chapter, we will look at how power supplies operate, and we will discuss band-pass, high-pass and low-pass filter circuits. Finally, we'll show you block diagrams of complete transmitters and receivers, and investigate how the stages are interconnected to make them work. You will be directed to turn to Chapter 10 at appropriate points, and to use the FCC questions as a study aid to review your understanding of the material.

Keep in mind that there have been entire books written on each topic covered in this chapter. If you do not understand some of the circuits from our brief discussion, it would be a good idea to consult some other reference books. *The ARRL Handbook for the Radio Amateur* is a good starting point, but even that won't tell you everything about a topic. The discussion in this chapter should help you understand the circuits well enough to pass your Technician/General class license exam, however.

POWER SUPPLIES

One circuit that is common to just about every type of radio circuit is the *power supply.* The 117-volt ac current available at your wall outlet isn't exactly what we mean by a power supply. The current at the outlet is an *alternating current,* and most electronic components require *direct current.* Additionally, circuits may require voltage levels other than the 117 volts available at the wall outlet; a power supply is used to increase or decrease the value of the voltage and to convert ac to dc.

An ac power supply consists of three sections: the transformer, the diode *rectifier* and the *filter.* A power transformer is used to raise or lower the voltage; it transforms the level of input voltage (117 or 234 V) to the level required by the equipment. Power transformers normally operate at 60 hertz, the frequency of household current.

Many power transformers have multiple secondary windings to provide several different output voltages. The power transformer used in a vacuum-tube amateur transmitter might have three secondaries: one delivering 6.3 volts for the tube filaments, one 12-volt winding for the solid-state circuits and one supplying high voltage for the plates of the final-amplifier tubes.

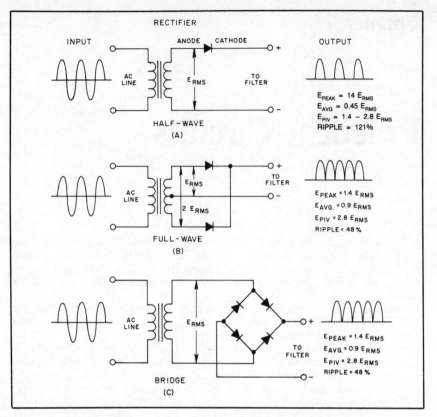

Figure 7-1—Fundamental rectifier circuits are shown. (A) Half-wave. (B) Full-wave center-tap. (C) Full-wave bridge. These rectifier circuits are discussed in the text.

Rectifiers

The diode rectifier section of a power supply converts ac to dc. (Remember that diodes conduct only in one direction.) Several common rectifier circuits are shown in Figure 7-1. Let's examine the operation of the simplest type, the *half-wave rectifier.*

To understand diode-rectifier action, we'll consider a diode connected to a secondary winding of a power transformer, as shown in Figure 7-1A. During the half of the ac cycle that the applied voltage makes the anode positive with respect to the cathode, the diode will conduct. Current can flow through the diode and any load attached to the output of the rectifier. The current passing through the load varies with time, as the diode will allow current to pass only during one half of the cycle. During the other half of the cycle, the anode is not positive with respect to the cathode, so no current passes through the diode. The action of the diode in permitting the current to flow only in one direction is called rectification.

The output of the rectifier is also shown in Figure 7-1A. The half-wave rectifier circuit takes its name from the fact that only half the ac wave passes through the circuit. One complete cycle of an ac wave goes from zero to 360 degrees and the half-wave rectifier uses only 180 degrees (one half) of the wave.

If another diode is connected to the opposite end of the transformer secondary, the second diode will conduct during the portions of the ac cycle that the first diode cannot. A connection (called a center tap) is made in the center of the transformer winding and connected through the load to the cathodes of the diodes. The rectifier circuit that has diodes working on both halves of the cycle is called a *full-wave rectifier*. This particular full-wave rectifier is known as a full-wave, center-tap rectifier. The advantage of the full-wave rectifier can be seen from the output waveform in Figure 7-1B—output is produced during the entire 360° of the ac cycle.

A third type of rectifier circuit is the *full-wave bridge rectifier* shown in Figure 7-1C. The output current or voltage waveshape is the same as for the full-wave center-tap rectifier. The important difference is in the output voltage from the rectifier compared to the voltage output from the secondary of the transformer. In the center-tap circuit, each side of the center-tapped secondary must develop enough voltage to supply the desired dc output voltage so the total transformer output voltage must be twice the power-supply output voltage (for a 6.3-V rectifier output we must use a 12.6-V center-tapped transformer). In the bridge circuit, no center tap is necessary—we can get 12.6-V output from a 12.6-V transformer. We can't get something for nothing, however—the elimination of the center tap requires the use of two additional rectifiers in the bridge-rectifier circuit. The bridge is another full-wave rectifier, and it uses the full 360 degrees of the ac wave.

In the bridge circuit, two rectifiers are connected to the transformer secondary in the same way as the two rectifiers in the center-tap circuit shown in Figure 7-1B. The second pair of rectifiers is also connected in series between the ends of the transformer winding, but in reverse. When the upper end of the transformer secondary is positive, the current path is through one rectifier, the load, and through another rectifier on its way back to the other end of the secondary. During the other half of the ac cycle, the current passes through the other two rectifiers. See Figure 7-2.

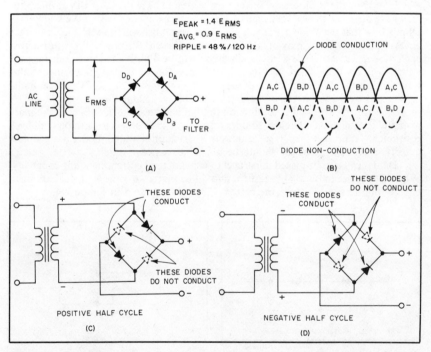

Figure 7-2—The operation of a full-wave, bridge-rectifier circuit is shown. The basic circuit is illustrated at A. Diodes A and C conduct on one-half of the input cycle, while diodes B and D conduct on the other half. The nonconducting diodes have the full transformer output voltage across them in the reverse direction.

Diode Ratings

Remember that diodes are rated as to their current-handling capability and maximum voltage that may be applied in the reverse direction. These ratings must not be exceeded. For example, when one diode in a bridge rectifier is conducting, the two diodes connected to it are not. These diodes must withstand the full transformer secondary voltage in the reverse direction. Since the peak value of the voltage is 1.41 times the RMS value, the diodes must withstand a peak inverse voltage of 1.41 times the RMS value of the transformer secondary. A safe design practice is to use diodes rated at twice the transformer output voltage.

In full-wave rectifier circuits, each diode carries one-half the output current, so the diodes must be rated to pass at least this much current without becoming too warm. Using diodes able to handle the full current (instead of half) is a safer design practice. In addition, when a capacitor-input filter is used, the capacitor appears as a dead short to the rectifier when the supply is first switched on. This means that the diodes may pass as much as 10 to 100 times the average current in one brief burst. The diodes used with a capacitor-input filter must be able to handle this surge current.

Diodes in Series and Parallel

Let's say we're building a power supply, and we need rectifiers rated at 1000 PIV, but we only have 500-PIV diodes on hand. How can we use the 500-PIV diodes? If we connect two 500-PIV diodes in series, the pair will withstand 1000 PIV (2 × 500 = 1000). To equalize the voltage drops and guard against transient voltage spikes, a resistor and a capacitor should be connected in parallel with each diode. See Figure 7-3.

Even though the diodes are of the same type and have the same PIV rating, they may have different "back" resistances when they are not conducting. We know from Ohm's Law that the diode with the higher back resistance will have the higher voltage across it, and the diode may break down. If we place a "swamping" resistor across each diode, the resistance across each diode will be almost the same and the back voltage will divide equally. A good rule of thumb for determining the resistor value is to multiply the PIV rating of the diode by 500. For example, a 500-PIV diode should be shunted by a 250-kilohm resistor (500 × 500 = 250,000).

The shift from forward conduction to high back resistance does not take place instantly in a silicon diode. Some diodes take longer than others to develop high back resistance. To protect the "fast" diodes in a string until all the diodes are properly cut off, a 0.01-μF capacitor should be placed across each diode.

Diodes can also be placed in parallel to increase total current-handling capability. When diodes are in parallel, a resistor should be placed in series with each diode to ensure that each takes the same current. This is illustrated in Figure 7-4.

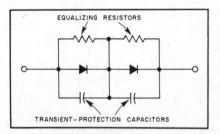

Figure 7-3—When diodes are used in series, equalizing resistors and transient-protection capacitors should be connected in parallel with each diode (see text).

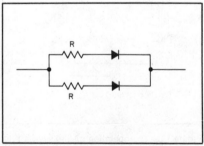

Figure 7-4—Use equalizing resistors when connecting diodes in parallel.

Power-Supply Filters

The output from a rectifier is not like dc from a battery. It's neither pure nor constant. It's actually what's known as *pulsating dc*—flowing only in one direction, but not at a steady value. Using this type of current as a replacement for a battery in a transmitter or receiver would introduce a strong hum on all signals. To smooth out the pulsations (called *ripple*) and eliminate the hum and its effects, a *filter* is used between the rectifier and the load.

The simplest type of filter is a large capacitor across a load (Figure 7-5). If the capacitance is high enough and the current drawn by the load is not excessive, the

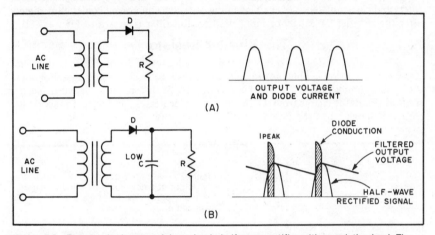

Figure 7-5—The circuit shown at A is a simple half-wave rectifier with a resistive load. The waveform shown to the right shows output voltage and diode current for this circuit. B illustrates how diode current and output voltage change with the addition of a capacitor filter. The diode conducts only when the rectified voltage is greater than the voltage stored in the capacitor. Since this time is usually only a short portion of a cycle, the peak current will be quite high.

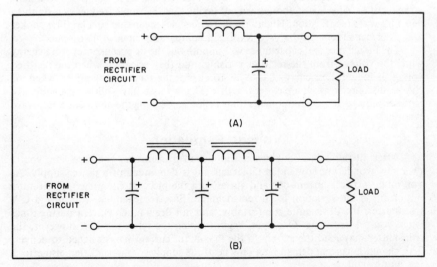

Figure 7-6—Shown at A is a choke-input power-supply filter circuit. B shows a capacitor-input, multisection filter.

capacitor will charge during voltage peaks. The capacitor discharges during the reversal and smooths out the ripple in the dc. During the peaks in the pulsating dc waveform, the capacitor stores energy. As the value of the dc waveform decreases, the capacitor releases some of its energy, and supplies current to the load. This cycle occurs over and over, resulting in a smooth dc output waveform, with minimum ripple.

We can design a more effective filter if we place an inductor, or *filter choke,* between the rectifier and the load, as shown in Figure 7-6A. The series inductor opposes changes in the circuit current the way the capacitor opposes changes in circuit voltage. We can connect several series-choke/parallel-capacitor sections one after another to improve the filtering action even more. The two basic filters are called *choke-input filters* and *capacitor-input filters,* depending on which component comes immediately after the rectifier. A capacitor-input multisection filter is shown in Figure 7-6B.

The Bleeder Resistor

It's a good design practice to connect a high-value resistor across the filter capacitor in a power supply, as shown in Figure 7-7, to discharge the capacitor when the power-supply primary voltage is turned off. Such a resistor, called a *bleeder resistor,* drains the charge from the capacitors and decreases the chances for accidental electrical shock when the power supply is turned off.

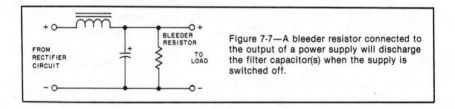

Figure 7-7—A bleeder resistor connected to the output of a power supply will discharge the filter capacitor(s) when the supply is switched off.

Ripple

Notice that in Figure 7-1 there are twice as many "bumps" in the dc output voltage per cycle of line frequency when a full-wave rectifier is used as compared to a half-wave rectifier. This is just another way of saying that the ripple frequency is doubled with full-wave rectification. Higher ripple frequencies mean that smaller filter chokes and filter capacitors can be used to obtain comparably smooth dc outputs.

The inductance and capacitance will smooth out the dc because they store energy while the voltage from the rectifier is rising, and they release it when the rectified voltage is falling. Less energy has to be stored, for the same overall effect, when the charge/discharge periods are closer together. That's why a filter following a half-wave rectifier requires larger components than a filter smoothing the output of a full-wave rectifier.

Voltage Regulation

After all this talk about power supplies in transmitters and receivers, you're probably wondering how much direct current is demanded of a power supply. As you might expect, current demand varies with the piece of equipment. Even more important is the variation in current demand. Say, for example, you have a CW transmitter. It will sit quite comfortably, and not draw much current during times when you are receiving. Once you begin tapping the telegraph key, however, the transmitter may load the supply to the maximum current it is designed to deliver. This change in current demand not only heats the supply components, but also affects the output voltage.

The voltage regulation of a power supply is a measure of the change in the output voltage with respect to the change in current. It is usually expressed as a percentage of the output voltage at the rated load current:

$$\text{Voltage Regulation} = \frac{V_n - V_1}{V_1} \times 100\% \qquad \text{(Equation 7-1)}$$

where
  V_n = voltage with no load
  V_1 = voltage under load

For example, a supply may be designed to deliver a current of 200 mA at 600 volts. If the voltage rises to 800 volts when the current is zero, the change is 200 volts. The regulation is:

$$\text{Voltage Regulation} = \frac{200}{600} \times 100\% = 0.333 \times 100\%$$

$$\text{Voltage Regulation} = 33\frac{1}{3}\%$$

One factor contributing to the voltage drop in the supply is the resistance and reactance of the transformer windings. Another is the voltage drop across the rectifier. The rectifier drop amounts to a fraction of a volt or so with silicon rectifiers, but can be significant if you use a tube rectifier—sometimes as high as 40 to 50 volts at full-rated current. A third factor is the resistance of the filter choke and any resistors that may be used for filtering. In all of these, the voltage drops increase with current. Finally, in a capacitor-input filter, the output voltage depends on the amount of energy stored in the input capacitor and how quickly the energy is released to the load. As the load current increases, so does the release rate, with a resulting voltage decrease. When there's no output current, the voltage in a capacitor-input filter builds up to the peak value of the rectifier voltages, or 1.41 times the RMS voltage applied to the rectifier.

[If you are studying for an Element 3B (General) exam, turn to Chapter 11 and study the questions in subelement 3G that begin 3G-1.]

FILTERS

We've already discussed one type of filter used in Amateur Radio, the power-supply filter. You should also be familiar with three others: the high-pass filter, the low-pass filter and the band-pass filter.

Low-Pass Filters

A *low-pass filter* is one that passes all frequencies below a certain frequency, called the *cutoff frequency*. Frequencies above the cutoff frequency are *attenuated*, or significantly reduced in amplitude. See Figure 7-8A. The cutoff frequency depends on the design of the low-pass filter. One other important trait that must be considered in the design of a low-pass filter is the characteristic impedance. The low-pass filter is always connected between the transmitter and the antenna, preferably as close to the transmitter as possible. The filter should have the same impedance as the feed line connecting the transmitter to the antenna, and filters should be used only in a feed line with a low standing-wave ratio. The cutoff frequency has to be higher than the highest frequency used for transmitting. A filter with a 45-MHz cutoff frequency would be fine for the "low bands" (1.8 through 29.7 MHz), but the filter would significantly attenuate 6-meter (50 MHz) signals.

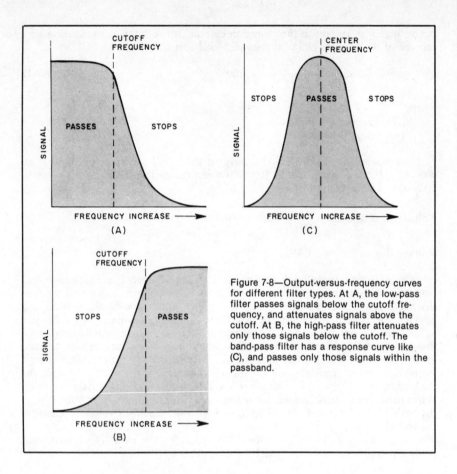

Figure 7-8—Output-versus-frequency curves for different filter types. At A, the low-pass filter passes signals below the cutoff frequency, and attenuates signals above the cutoff. At B, the high-pass filter attenuates only those signals below the cutoff. The band-pass filter has a response curve like (C), and passes only those signals within the passband.

High-Pass Filters

A *high-pass filter* passes all frequencies above the cutoff frequency, and attenuates those below it. See Figure 7-8B. A high-pass filter connected to a television set, stereo receiver or other home-entertainment device that is being interfered with will attenuate the signal from an amateur station while allowing the higher-frequency television or FM broadcast signals to pass through to the receiver. This is useful in reducing the amount of lower-frequency radiation (at the ham's operating frequency) that might overload a television set, causing disruption of reception. For best effect, the filter should be connected as close to the television set's tuner as possible—inside the TV if it is your own set and you don't mind opening it up, or directly to the antenna terminals on the back of the television. If the set belongs to your neighbors, have them contact a qualified service technician to install the filter, so you are not blamed if something goes wrong with the set later on.

Band-Pass Filters

A *band-pass filter* is a combination of a high-pass and low-pass filter. It passes a desired range of frequencies while rejecting signals above and below the pass band. This is shown in Figure 7-8C. Band-pass filters are commonly used in receivers to provide different degrees of rejection—very narrow filters are used for CW reception, wider filters for single-sideband voice (SSB) and AM double-sideband reception.

[If you are studying for an Element 3A (Technician) exam, turn to Chapter 10 and study questions 3G-2.1 through 3G-2.7 and question 3G-2.9.

If you are studying for an Element 3B (General) exam, study question 3G-2.8 in Chapter 11.]

THE RECEIVER

The fundamental function of any radio receiver is to change radio-frequency signals (which we can't hear) to audio-frequency signals (which we can). A good receiver has the ability to find weak radio signals, to separate them from other signals and interference, and to stay tuned to one frequency without drifting.

The ability of a receiver to find or detect weak signals is called *sensitivity*. *Selectivity* is the ability to separate (select) a desired signal from undesired signals. *Stability* is a measure of the ability of a receiver to continue receiving one particular frequency. In general, then, a good receiver is very sensitive, selective and stable.

Detection

The basic duty of a receiver is to extract the audio information from a radio signal. To detect means to discover, and the *detector* stage in a receiver is where the discovery (or recovery) of the audio information takes place.

One simple receiver is shown in Figure 7-9A. This receiver is called a crystal receiver because it uses a simple crystal diode as the detector. At B in Figure 7-9 is

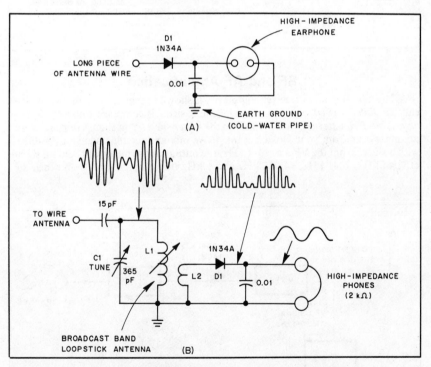

Figure 7-9—Illustration A shows a simple crystal AM receiver. It consists only of a wire antenna, detector diode, capacitor, earphone and earth ground. The circuit at B has a tuned circuit that helps separate the broadcast-band signals, but otherwise operates in the same manner as the circuit at A.

a more complex crystal receiver. An incoming radio signal causes a current to flow from the antenna to ground. The current induces a voltage with the same waveform in L2 when the L1-C1 circuit is resonant at the frequency of the incoming radio signal. The diode rectifies the RF signal, allowing only half the wave to pass through. Capacitor C2 bypasses the RF portion of the signal to ground, leaving only the audio signal to pass on to the headphones.

The detector is really the heart of a receiver, and everything else is sort of "extra equipment." In the early days of radio, crystal receivers using galena crystals and "cat's whiskers" were in common use. See Figure 7-10. On today's crowded frequencies, however, more sensitive and selective receivers are required. In addition, crystal receivers cannot be used to receive SSB and CW signals properly. The crystal receiver serves to show how simple a receiver can be, and is an excellent example of how detection is accomplished. Every receiver will have some type of detector stage.

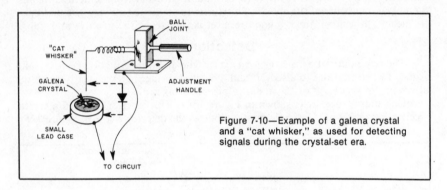

Figure 7-10—Example of a galena crystal and a "cat whisker," as used for detecting signals during the crystal-set era.

RF and AF Amplification

The next step up in receiver complexity is shown in Figure 7-11. In this receiver, called a *direct-conversion receiver,* the incoming signal is combined with a signal from a *variable-frequency oscillator (VFO)* in the *mixer* stage. You should remember how mixing works from your Novice exam. If we mix an incoming signal at 7040 kHz with a signal from the VFO at 7041 kHz, the output of the mixer will contain signals at 7040 kHz, 7041 kHz, 14,081 kHz and 1 kHz (the original signals and their sum

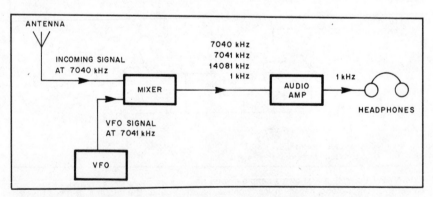

Figure 7-11—Block diagram of a direct-conversion receiver. This type of receiver converts RF signals directly to audio, using only one mixer.

and difference frequencies). One of these signals (1 kHz) is low enough in frequency that we can amplify it with an audio amplifier and hear it in headphones.

This receiver is called a direct-conversion receiver because the RF signal is converted directly to audio, in one step. By changing the VFO frequency, other signals in the range of the receiver can be converted to audio signals. The bandwidth of a tuned circuit increases as the operating frequency increases. This can create serious selectivity problems in a simple receiver designed for the higher-frequency bands, and is a major limitation of direct-conversion receivers.

Superheterodyne Receivers

A block diagram of a simple *superheterodyne receiver* for CW and SSB is shown in Figure 7-12. The first mixer is used to produce a signal at the *intermediate frequency*

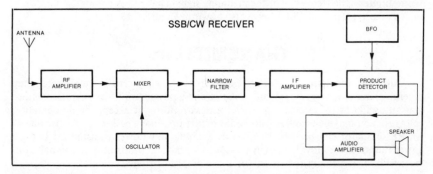

Figure 7-12—Block diagram of a superheterodyne SSB/CW receiver. This type of receiver uses one mixer to convert the incoming signal to the intermediate frequency (IF) and another mixer, called a product detector, to recover the audio or Morse code.

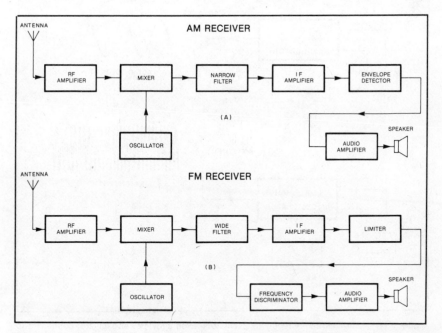

Figure 7-13—AM and FM receivers use similar stages, with a few variations.

or IF, typically 9 MHz in a modern receiver. All the amplifiers after this first mixer stage are designed for peak efficiency at the IF. The superhet receiver solves the selectivity/bandwidth problem by converting all signals to the same IF before filtering and amplification. To receive SSB and CW signals, a second mixer, called a product detector, is used. The product detector mixes the IF with a signal from the *beat-frequency oscillator* (BFO). The output from the product detector contains audio components that can be amplified and sent to a speaker or headphones.

An FM receiver is organized along the same lines, with a few different stages, such as a wider bandwidth filter and a different type of detector, called a frequency discriminator. Also, most FM receivers have a limiter stage added between the IF amplifier and the detector. The limiter chops off noise and amplitude variations from an incoming signal, and the frequency discriminator produces a waveform varying in amplitude as the frequency of the incoming signal changes. A comparison between an AM receiver and an FM receiver is shown in Figure 7-13.

TRANSMITTERS

Although amateur transmitters range from the simple to the elaborate, they generally have two basic stages, an *oscillator* and a power amplifier (PA). Remember that an oscillator produces an ac waveform with no input except the dc operating voltages. You may be familiar with audio oscillators, like a code-practice oscillator. An RF oscillator can be used by itself as a simple low-power transmitter, but the power output is very low. See Figure 7-14. In a practical transmitter the signal from the oscillator is usually fed to one or more amplifier stages.

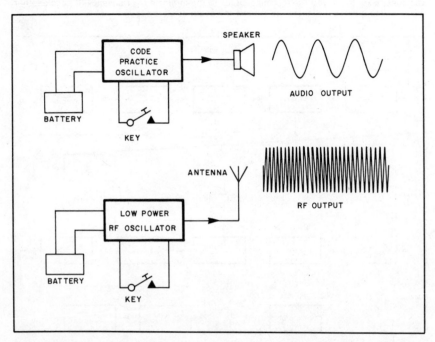

Figure 7-14—Audio- and radio-frequency oscillators. Even a low-power oscillator connected to a good antenna can send signals hundreds of miles.

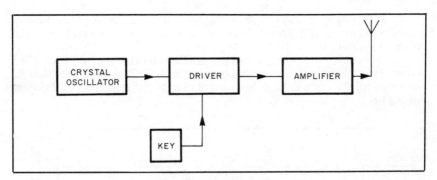

Figure 7-15—Block diagram of a basic CW transmitter.

A block diagram of a simple amateur transmitter is shown in Figure 7-15. The simple transmitter consists of a crystal oscillator followed by a driver stage and a power amplifier. A crystal oscillator uses a quartz crystal to keep the frequency of the radio signal constant.

Crystal oscillators are not practical in many cases. A different crystal is needed for each desired operating frequency. After a while, this becomes quite expensive as well as impractical. If we use a variable-frequency oscillator (VFO) in place of the crystal oscillator, we can change the transmitter frequency at will. Great care in design and construction is required if the stability of a VFO is to compare with that of a crystal oscillator. Stability means the same thing in transmitter design as it does in receiver design; it is the property of a transmitter to remain on one frequency without drifting.

The block diagram in Figure 7-16 shows a typical AM transmitter containing a VFO. Immediately following the VFO is a stage called a buffer. The buffer may amplify the signal slightly (and is sometimes called a buffer amplifier), but its primary purpose is to isolate the VFO from the following amplifier stages. Without the buffer, the final amplifiers might load the VFO to the point that it would be pulled off frequency. The resulting signal would have a pronounced chirpy sound in a receiver.

The buffer amplifier is designed to keep a constant load on the oscillator. Occasionally, we use the buffer to amplify the oscillator output so there's enough power to drive the power amplifier. Buffer circuits provide isolation between the output (plate or collector) and the input (grid or base).

A *frequency multiplier* produces an output at a multiple of the input frequency. It is an amplifier with its output tuned to the desired multiple (harmonic) of the input signal. A doubler is a multiplier that gives output at twice the input frequency; a tripler multiplies the input frequency by three, and so on.

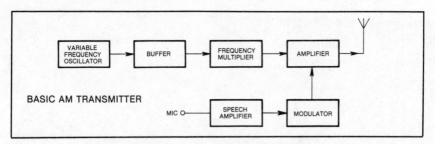

Figure 7-16—In an AM double-sideband voice transmitter, the modulation is usually added in the final-amplifier section.

From the viewpoint of any particular stage in a transmitter, the stage that precedes it is called its driver. In Figure 7-16, the buffer is the driver of the multiplier. The multiplier, in turn, drives the power amplifier.

Voice modulation is usually added to the RF signal in the amplifier stage of an AM double-sideband voice transmitter, as shown in Figure 7-16. For single-sideband transmissions, the modulating audio is added in an earlier stage, called a *balanced modulator*.

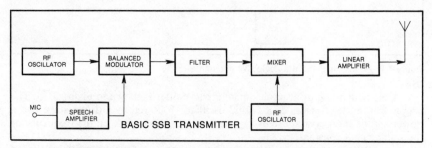

Figure 7-17—In an SSB transmitter, the modulation is added and the carrier removed by a circuit called a balanced modulator. After this stage, the signal is mixed with another signal to reach the desired operating frequency.

The balanced modulator also balances out (cancels) the original carrier signal. We are left with a double-sideband, suppressed-carrier signal. The filter in Figure 7-17 is used to filter out one of the sidebands, leaving us with an SSB signal. Next, the SSB signal is mixed with a tunable RF signal to convert it to our desired operating frequency. The output of the mixer stage is then amplified and fed into the antenna.

Following the generation of a single-sideband signal, its frequency can be changed only by frequency conversion, not by multiplication. For this reason, the heterodyning exciter shown in Figure 7-17 is almost always used in SSB transmitters.

In transmitters of this type, all of the circuits, except for the final amplifier, operate at low signal levels, similar to those in a receiver. It is very practical to use transistors for these low-level stages. Typically, vacuum tubes are now reserved for use at higher power levels, such as external power amplifiers operating at the kilowatt level.

The most significant advantage of the conversion (heterodyne) exciter is that some of the same components and circuits can be used for both transmitting and receiving. A combination transmitting/receiving setup is called a transceiver (transmitter and receiver in one unit).

[If you are studying for an Element 3A (Technician) exam, turn to Chapter 10 and study the questions in subelement 3G that begin 3G-3.]

Key Words

Amateur Teleprinting Over Radio (AMTOR)—a modified ITA2 (International Telegraphic Alphabet Number 2) code that has error-correcting capabilities

American National Standard Code for Information Interchange (ASCII)—a seven-bit code used for radioteleprinter and computer applications

Amplitude modulation—a method of combining an information signal and an RF carrier wave in which the amplitude of the RF envelope (carrier and sidebands) is varied in relation to the information signal strength

Audio-frequency shift keying (AFSK)—a method of transmitting radioteletype information by switching between two audio tones fed into the transmitter microphone input. This is the RTTY mode most often used on VHF and UHF.

Balanced modulator—a circuit used in a single-sideband suppressed-carrier transmitter to combine a voice signal and the RF signal. The balanced modulator isolates the input signals from each other and the output, so that only the difference of the two input signals reaches the output.

Bandwidth—the frequency range (measured in hertz) over which a signal is stronger than some specified amount below the peak signal level. For example, if a certain signal is at least half as strong as the peak power level over a range of ± 3 kHz, the signal has a 3-dB bandwidth of 6 kHz.

Baudot code—the term used for ITA2 in the US

Deviation ratio—the ratio between the maximum change in RF-carrier frequency and the highest modulating frequency used in an FM transmitter

Emission designator—a symbol made up of two letters and a number, used to describe a radio signal. A3E is the designator for double-sideband, full-carrier, amplitude-modulated telephony.

Frequency deviation—the amount the carrier frequency in an FM transmitter changes as it is modulated

Frequency modulation—the process of varying the frequency of an RF carrier in response to the instantaneous changes in an audio signal

Frequency-shift keying (FSK)—a method of transmitting radioteletype information by switching an RF carrier between two separate frequencies. This is the RTTY mode most often used on HF.

International Telegraphic Alphabet Number 2 (ITA2)—a five-bit code used for radioteleprinter transmissions

Modulation—the process of varying some characteristic (amplitude, frequency or phase) of an RF carrier for the purpose of conveying information

Modulation index—the ratio between the maximum carrier frequency deviation and the audio modulating frequency at a given instant in an FM transmitter

Peak envelope power (PEP)—the average power of the RF cycle having the greatest amplitude. (This occurs during a modulation peak.)

Phase—the time interval between one event and another in a regularly recurring cycle

Phase modulation—varying the phase of an RF carrier in response to the instantaneous changes in an audio signal

Reactance modulator—an electronic circuit whose capacitance or inductance changes in response to an audio input signal

RF envelope—the shape of an RF signal as viewed on an oscilloscope

Sideband—the sum or difference frequencies generated when an RF carrier is mixed with an audio signal

Single-sideband, suppressed-carrier, amplitude modulation (SSB)—a technique used to transmit voice information in which the amplitude of the RF carrier is modulated by the audio input, and the carrier and one sideband are suppressed

Splatter—interference to adjacent signals caused by overmodulation of a transmitter

Varactor diode—a component whose capacitance varies as the reverse-bias voltage is changed

Chapter 8

Signals and Emissions

O ne of the most exciting aspects of Amateur Radio is that it offers so many different ways to participate. As a Novice, you were limited to CW, also known as A1A emission. As a Technician or General you'll be allowed to use several modes or emission types. This chapter will introduce the modes and emission types you will need to understand for your FCC exam.

FCC EMISSION DESIGNATORS

The FCC uses a special system to specify the types of signals (emissions) permitted to amateurs and other users of the radio spectrum. Each designator has three digits; Table 8-1 shows what each character in the designator stands for. The

---

Table 8-1

Partial List of WARC-79 Emission Designators

(1) First Symbol — Modulation Type

Unmodulated carrier	N
Double sideband full carrier	A
Single sideband reduced carrier	R
Single sideband suppressed carrier	J
Vestigial sideband	C
Frequency modulation	F
Phase modulation	G
Various forms of pulse modulation	P, K, L, M, Q, V, W, X

(2) Second Symbol — Nature of Modulating Signals

No modulating signal	0
A single channel containing quantized or digital information without the use of a modulating subcarrier	1
A single channel containing quantized or digital information with the use of a modulating subcarrier	2
A single channel containing analog information	3
Two or more channels containing quantized or digital information	7
Two or more channels containing analog information	8

(3) Third Symbol — Type of Transmitted Information

No information transmitted	N
Telegraphy — for aural reception	A
Telegraphy — for automatic reception	B
Facsimile	C
Data transmission, telemetry, telecommand	D
Telephony	E
Television	F

---

designators begin with a letter that tells what type of modulation is being used. The second character is a number that describes the signal used to modulate the carrier, and the third character specifies the type of information being transmitted.

Some of the more common combinations are:

- NØN—Unmodulated carrier
- A1A—Morse code telegraphy using amplitude modulation
- A3E—Double-sideband, full-carrier, amplitude-modulated telephony
- J3E—Amplitude-modulated, single-sideband telephony with suppressed carrier
- F3E—Frequency-modulated telephony
- F1B—Telegraphy using frequency-shift keying without a modulating audio tone (FSK RTTY). F1B is designed for automatic reception.
- F2A—Telegraphy produced by the on-off keying of an audio tone fed into an FM transmitter.
- F2B—Telegraphy produced by modulating an FM transmitter with audio tones (AFSK RTTY). F2B is also designed for automatic reception.
- G3E—Phase-modulated telephony

Signal Quality

The quality of the signal your station produces reflects on you! You, as a radio amateur and control operator, are responsible (legally and otherwise) for all the signals emanating from your station. Your signals are the only way an amateur on the other side of the country or the world will get to know you.

Communication is the transfer of information from one place to another. We can derive information from a radio signal just by its presence and its strength, as in the case of a beacon. By switching the signal on and off in some prescribed manner we can transmit more information. Finally, we can modify the signal in other ways to convey still more information. An important consideration and limitation, however, is the amount of space in the radio-frequency spectrum that a signal occupies. This space is called *bandwidth*. The bandwidth of a transmission is determined by the information rate. Thus, a pure, continuous, unmodulated carrier has a very small bandwidth, while a television transmission, which contains a great deal of information, is several megahertz wide.

[If you are studying for an Element 3A (Technician) exam, turn to Chapter 10 and study the questions in subelement 3H that begin 3H-1. Review this section if you have problems.]

MODULATION

Modulation is the process of varying some characteristic (amplitude, frequency or phase) of an RF carrier wave in accordance with the instantaneous variations of some external signal, for the purpose of conveying information. Almost every amateur will think of phone transmission when the word modulation is mentioned. The subject of modulation, however, is much broader than that. The principles are the same whether the radio transmission is modulated by speech, audio tones or the on-off keying that modulates a CW transmitter with Morse code.

The sound of your voice consists of physical vibrations in the air, called sound waves. The range of audio frequencies generated by a person's voice may be quite large; the normal range of human hearing is from 20 Hz to 20,000 Hz. All the information necessary to make your voice understood, however, is contained in a narrower band of frequencies. In amateur communications, your audio signals are usually filtered so that only frequencies between about 300 and 3000 hertz are transmitted. This filtering is done to reduce the bandwidth of the phone signal.

Microphones

The first step in phone transmission is to convert the sound of your voice into electrical signals that contain the same information. The device that does this is the microphone. Several types are commonly used by amateurs.

Piezoelectric microphones make use of the phenomenon by which certain materials produce a voltage through mechanical stress or distortion of the material. A diaphragm is physically coupled to a small bar of material such as Rochelle salt or ceramic. The motion of the diaphragm stresses the material, producing a very small voltage that varies in relation to the sound waves. A microphone containing Rochelle salt is called a crystal microphone, while one containing ceramic material is known as a ceramic microphone. These microphones generate very small signals. Their frequency response depends on the resistance of the load the microphone operates into.

Another common microphone in use is the dynamic microphone, which consists of a lightweight coil suspended in a magnetic field. The coil is attached to a diaphragm. When sound waves hit the diaphragm, the coil moves in the magnetic field and generates an alternating current. Yet another type, the electret microphone, uses a set of biased plates similar to a capacitor. These plates are set in motion by air-pressure variations, which create a changing capacitance and an accompanying change in output voltage.

Amplitude Modulation

On the high-frequency amateur bands, *single-sideband, suppressed-carrier, amplitude modulation* (usually called *SSB*) is the most widely used radiotelephone mode. Since SSB is a sophisticated form of *amplitude modulation,* it is worthwhile to take a look at some fundamentals of amplitude modulation.

Modulation is a mixing process. When RF and AF signals are combined in a standard double-sideband, full-carrier, AM (emission A3E) transmitter (such as one used for commercial broadcasting) four output signals are produced: the original carrier (RF signal), the original AF signal, and two *sidebands,* whose frequencies are the sum and difference of the original RF and AF signals. See Figure 8-1. The amplitude of the sidebands varies with the amplitude of the audio signal.

For higher-frequency modulating tones, the sideband frequencies will be further from the carrier frequency, and for lower-frequency tones the sidebands will be closer to the carrier frequency. The sum component is called the upper sideband, and increasing the frequency of the modulating audio causes a corresponding increase in the frequency of the RF output signal. The difference frequency is called the lower sideband, and is inverted, meaning that an increase in the frequency of the modulating AF causes a decrease in the frequency of the output signal. The amplitude and frequency of the carrier are unchanged by the modulation process.

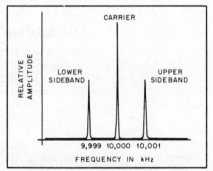

Figure 8-1—A 10-MHz carrier modulated by a 1-kHz sine wave has sidebands at 9.999 MHz and 10.001 MHz.

The resulting *RF envelope* (the sum of the sidebands and carrier) as viewed on an oscilloscope, has the shape of the modulating waveform, as shown in Figure 8-2. The modulating waveform is mirrored above and below the zero axis of the RF envelope.

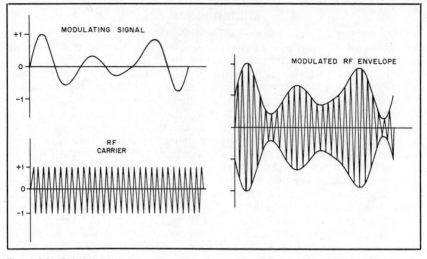

Figure 8-2—Relationship between the modulating audio waveform, the carrier and the resulting RF envelope in a double-sideband, full-carrier AM signal.

The amount of spectrum space occupied by the sidebands in a properly operating A3E transmitter depends on the bandwidth of the audio modulating signal. This is why most amateur transmitters limit the maximum audio-frequency component of your voice that is transmitted to about 3000 Hz. There is one emission type that does not produce sideband frequencies. That emission, designated NØN, is an unmodulated carrier signal.

[If you are studying for an Element 3A (Technician) exam, turn to Chapter 10 and study questions 3H-2.2 and 3H-9.2.

If you are studying for an Element 3B (General) exam, study the following questions in Chapter 11: 3H-2.1, 3H-2.3, 3H-2.4, 3H-3.1, 3H-3.2 and 3H-11.1.]

Modulation Percentage

One of the concepts that is easy to visualize by looking at the RF envelope is percentage of modulation. Figure 8-3A is a sine-wave signal used to modulate an RF carrier. In Figure 8-3B, we see that the modulating signal has just the right value to make the RF-envelope amplitude go to zero on the audio downswing and to twice the unmodulated value at the peak of the upswing. This is called 100-percent modulation. The percent of modulation is the ratio of the peak-to-peak voltage of the modulating signal to the peak-to-peak voltage of the unmodulated RF envelope. With voice modulation, the percentage is continually varying, because speech contains many variations in volume. If the modulating signal were only half as strong, so its peak-to-peak voltage was 1 instead of 2 as shown in Figure 8-3A, the modulated envelope would look like the one shown in part C. This represents a condition of 50% modulation.

Occasionally the envelope may be held at zero for longer than an instant, as shown in Figure 8-3C. This produces a gap in the RF output that distorts the shape of the modulation envelope so that it differs greatly from the modulating signal. This situation is called overmodulation, since the carrier is modulated more than 100%. The distorted signal introduces harmonics and can cause "splatter"—a signal with excessive bandwidth that results in interference to nearby signals.

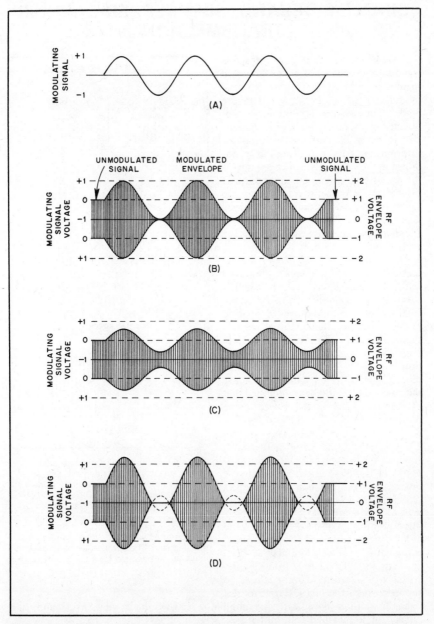

Figure 8-3—The audio-frequency waveform at A is superimposed on an RF signal, and the resulting RF envelope is shown by the outline at B. The RF carrier is 100% modulated—the envelope just barely reaches zero between peaks. The RF envelope shown at C is less than 100% modulated because the maximum amplitude of the audio signal was decreased from the amplitude shown at A. At D, the audio-signal strength was increased, and the RF envelope is modulated more than 100%.

Signals and Emissions   8-5

SINGLE-SIDEBAND, SUPPRESSED-CARRIER TRANSMISSION

We know that when the carrier signal is modulated, sidebands are generated in proportion to the strength of the modulating signal, as shown in Figure 8-4A. The carrier amplitude is not affected by the modulation. Since the carrier is never changed, it contains no information. All the information is contained in the sidebands, which are mirror images of each other. The only purpose of the carrier is to provide a reference signal for demodulation by an envelope detector.

Suppose we remove or suppress the carrier and amplify only the sidebands, as shown in Figure 8-4B. Removing the carrier means we can increase the power in the sidebands by using the power that was required to amplify the carrier in the final stages. The total power dissipation in the transmitter final amplifier remains the same. Taking this even further, since both sidebands contain the same information, we really need only one. If we remove one of the sidebands we can use the power that would have amplified that sideband to increase the power in the remaining sideband. See Figure 8-4C.

The FCC uses the emission symbol J3E to designate single-sideband, suppressed-carrier signals. J3E is the most popular form of voice modulation in use on the amateur HF bands. FCC rules require that the carrier be reduced to a level at least 40 dB below the peak effective power output of the desired sideband in a J3E transmitter. (This specification is found in FCC Docket number 80-739.)

You should remember from Chapter 7 that a balanced modulator is used to suppress the carrier in a J3E transmitter. A bandpass filter placed after the modulator removes the unwanted sideband. In a typical single-sideband transmitter a 2.5-kHz bandpass filter is used.

By now, you probably realize that by removing one of the sidebands and the carrier, the range of radio frequencies is reduced quite a bit as compared to a double-sideband, full-carrier transmission. In fact, SSB signals have the narrowest bandwidth of any of the popular telephony emissions.

[If you are studying for an Element 3B (General) exam, turn to Chapter 11 and study the following questions: 3H-4.1, 3H-4.2, 3H-5.1 and 3H-5.2.]

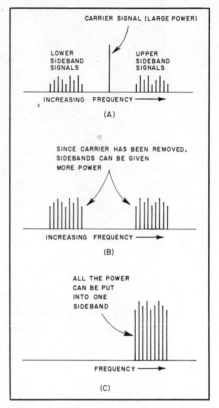

Figure 8-4—Amplitude-modulated radio signals. At A, a double-sideband, full-carrier signal. If the carrier is removed, as shown at B, the sidebands can be given more power without increasing the total transmitter power. Going further, if one sideband is suppressed, as shown at C, the full transmitter power can be concentrated in one sideband. The suppressed parts of the signal are not needed for intelligibility.

SSB Power Measurement

In a single-sideband, suppressed-carrier transmission, there is no signal when there is no modulation applied to the transmitter. When the transmitter is modulated, a constantly varying RF envelope is produced, like the one shown in Figure 8-5. Two amplitude values associated with the wave are of particular interest. One is the peak envelope voltage, the greatest amplitude reached by the envelope at any time. The other is the RMS voltage of the envelope, as read on an RMS-reading voltmeter with an RF probe. See Figure 8-6.

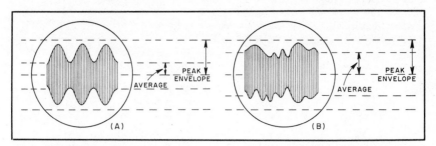

Figure 8-5—Two RF-envelope patterns that show the difference between average amplitude and peak amplitude. In B, the average amplitude has been increased, but the peak amplitude remains constant.

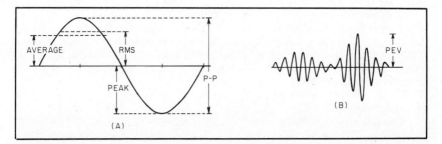

Figure 8-6—Ac voltage and current values used in making power measurements. The sine-wave parameters are illustrated at A, while B shows the peak envelope voltage (PEV) for a composite waveform.

The power measurement used to rate transmitter output is *peak envelope power (PEP)*. This is the average power of the wave at a modulation peak. One way to calculate PEP is to measure the peak envelope voltage, using an oscilloscope. Then multiply this value by 0.707 to calculate the RMS value of the peak envelope voltage. Remember that power equals current times voltage:

$$P = I \times E$$ (Equation 8-1)

We don't have to calculate current if we know the load impedance, however. For most amateur antenna systems we want a load impedance of about 50 ohms. From Ohm's Law, current equals voltage divided by resistance:

$$I = \frac{E}{R}$$ (Equation 8-2)

So we can substitute for I in the power equation:

$$P = \frac{E}{R} \times E = \frac{E^2}{R} \qquad \text{(Equation 8-3)}$$

Suppose we measured the output of a transmitter using an oscilloscope and found 250-V peak-to-peak across a 50-Ω dummy load. This represents a peak voltage of 125, and an RMS value of 88.4 V:

Peak envelope voltage = peak-to-peak voltage $\times$ 0.5
PEV = 250-V peak-to-peak $\times$ 0.5 = 125 V
RMS voltage = PEV $\times$ 0.707
RMS voltage = 125 V $\times$ 0.707 = 88.4 V

Using this value for E in Equation 8-3, we can calculate the PEP of our transmitter.

$$P = \frac{E^2}{R} = \frac{(88.4)^2}{50}$$

$$P = \frac{7810}{50} = 156 \text{ W}$$

Average power is found by using the value for RMS voltage as read on a voltmeter with an RF probe attached. Substitute that value for E in Equation 8-3.

For a constant-amplitude sine wave, such as for a CW signal, PEP and average power are the same. This is not true for a modulated wave.

Another way to measure the power of an amateur transmitter is with a peak-reading wattmeter. This type of power meter has capacitors in the metering circuit that charge to the peak voltage of the input wave. This voltage is then measured with a voltmeter and displayed on a scale calibrated in watts. To measure the actual transmitter power output, the wattmeter should be connected as close to the transmitter output terminal as possible.

[If you are studying for an Element 3A (Technician) exam, turn to Chapter 10 and study question 3D-5.2.

If you are studying for an Element 3B (General) exam, study the following questions in Chapter 11: 3D-5.1, 3D-5.5, 3D-5.6 and 3D-5.7. Also study these questions: 3H-8.1, 3H-8.2, 3H-9.1, 3H-13.1 and 3H-13.2.]

FREQUENCY MODULATION (FM)

We can transmit information by modulating any property of a carrier. We have already studied amplitude modulation; we can also modulate the frequency or the phase of the carrier. *Frequency modulation* and *phase modulation* are closely related, since the phase of a signal cannot be varied without also varying the frequency, and vice versa. Phase modulation and frequency modulation are especially suited for channelized local UHF and VHF communication because they feature good audio fidelity and high signal-to-noise ratio, as long as the signals are stronger than a minimum, or threshold, level.

In FM systems, when a modulating signal is applied, the carrier frequency is increased during one half of the audio cycle and decreased during the other half of the cycle. The change in the carrier frequency, or *frequency deviation,* is proportional to the instantaneous amplitude of the modulating signal. Figure 8-7 shows a representation of a frequency modulated signal. The deviation is slight when the amplitude of the modulating signal is small and greatest when the modulating signal reaches its peak. The amplitude of the envelope does not change with modulation.

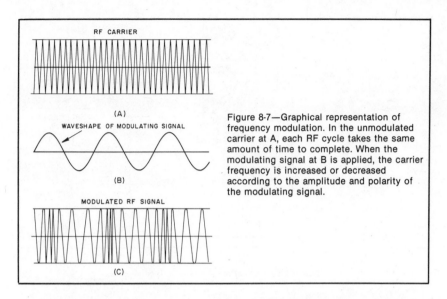

Figure 8-7—Graphical representation of frequency modulation. In the unmodulated carrier at A, each RF cycle takes the same amount of time to complete. When the modulating signal at B is applied, the carrier frequency is increased or decreased according to the amplitude and polarity of the modulating signal.

This means that an FM receiver is not sensitive to amplitude variations caused by impulse-type noise, and this feature makes it popular for mobile communications.

Practical FM Transmitters

The simplest type of FM transmitter we could design is shown in Figure 8-8. Remember that a circuit containing capacitance and inductance will be resonant at some frequency. If we use a resonant circuit in the feedback path of an oscillator, we can control the oscillator frequency by changing the resonant frequency of the tuned circuit. A capacitor microphone is nothing more than a capacitor with one movable plate (the diaphragm). When you speak into the microphone, the diaphragm vibrates and the spacing between the two plates of the capacitor changes. When the spacing changes, the capacitance value changes. If we connect the microphone to the resonant circuit in our oscillator, we can vary the oscillator frequency by speaking into the microphone. If we add frequency multipliers to bring the oscillator frequency up to our operating frequency, and amplifiers to bring up the power, we have a simple FM transmitter.

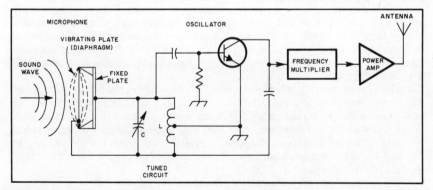

Figure 8-8—A simple FM transmitter. The frequency of the oscillator is changed by changing the capacitance in the resonant circuit.

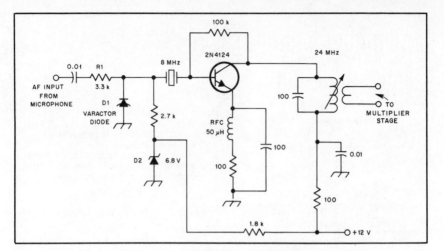

Figure 8-9—A varactor diode can also be used to frequency-modulate an oscillator. The capacitance of the varactor diode changes when the bias voltage is varied.

Another way to shift the frequency of the oscillator would be to use a circuit called a *reactance modulator*. This is a vacuum tube or transistor wired in a circuit so that it changes either the capacitance or inductance of the oscillator resonant circuit in response to an audio input signal.

Modern FM transmitters may use a *varactor diode* to modulate the oscillator. A varactor diode is a special diode that changes capacitance when its bias voltage is changed. Connected to a crystal oscillator as shown in Figure 8-9, the varactor diode will "pull" the oscillator frequency slightly when the audio input changes.

PHASE MODULATION (PM)

One problem with early direct-modulated FM transmitters was frequency stability. Using a capacitive microphone or a reactance modulator means that a crystal oscillator cannot be used. Frequency multiplication also multiplies any drift or other instability problems in the oscillator. With phase modulation, however, the modulation takes place after the oscillator, so a crystal oscillator can be used. For this reason phase modulation is used in most FM transmitters built for mobile applications, where stability may be a serious problem because of shocks and temperature variations. Phase modulation produces what is called indirect-modulated FM. The most common method of generating a phase-modulated telephony signal (emission designator G3E) is to use a reactance modulator.

The term *phase* essentially means "time," or the time interval between the instant when one thing occurs and the instant when a second related thing takes place. The later event is said to lag the earlier, while the one that occurs first is said to lead. When two waves of the same frequency start their cycles at slightly different times, the time difference (or phase difference) is measured in degrees. This is shown in Figure 8-10. In phase modulation, we shift the phase of the output wave in response to the audio input signal.

The output from an RF oscillator will be an unmodulated wave, as shown in Figure 8-11A. The phase-modulated wave is shown in Figure 8-11B. At time 0 and again 168 microseconds later the two waves are in phase. The phase modulation has

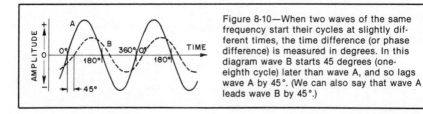

Figure 8-10—When two waves of the same frequency start their cycles at slightly different times, the time difference (or phase difference) is measured in degrees. In this diagram wave B starts 45 degrees (one-eighth cycle) later than wave A, and so lags wave A by 45°. (We can also say that wave A leads wave B by 45°.)

shifted cycle X′ so that it reaches its peak value at time 39 microseconds as compared with 52 microseconds for the unmodulated wave. (Cycle X′ leads cycle X.)

The number of cycles in both waveforms is the same, so the center frequency of the wave has not been changed. All of the cycles to the left of cycle X′ have been compressed, and all the cycles to the right of cycle X′ have been spread out. The frequency of the wave to the left is greater than the center frequency, and the frequency to the right is less—the wave has been frequency modulated.

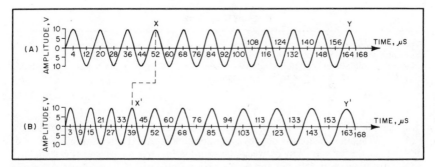

Figure 8-11—Graphical representation of phase modulation. The unmodulated wave is shown at A. After modulation, cycle X′ lags cycle X, and all the cycles to the left of cycle X′ are compressed. To the right, the cycles are spread out.

Frequency Deviation with Phase Modulation

With direct FM, frequency deviation is proportional to the amplitude of the audio modulating signal. With phase modulation, the frequency deviation is proportional to both the amplitude and the frequency of the modulating signal. Frequency deviation is greater for higher audio frequencies. This is actually a benefit, rather than a drawback, of phase modulation. In radiocommunications applications, speech frequencies between 300 and 3000 Hz need to be reproduced for good intelligibility. In the human voice, however, the natural amplitude of speech sounds between 2000 and 3000 Hz is low. Something must be done to increase the amplitude of these frequencies when a direct FM transmitter is used. A circuit called a preemphasis network amplifies the sounds between 2000 and 3000 Hz. With phase modulation, the preemphasis network is not required, because the deviation already increases with increasing audio input frequency.

Modulation Index

The sidebands that occur from FM or PM differ from those resulting from AM. With AM, only a single set of sidebands is produced for each modulating frequency. FM and PM sidebands occur at integral multiples of the modulating frequency on

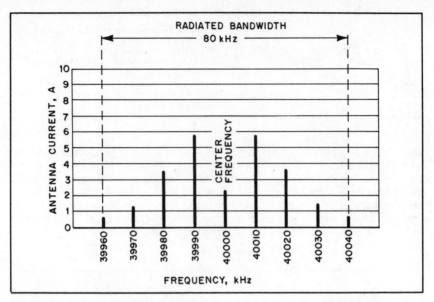

Figure 8-12—When an RF-carrier is frequency-modulated by an audio signal, multiple sidebands are produced. This diagram shows a 40-MHz carrier modulated by a 10-kHz sinusoidal audio signal. Additional sidebands exist above 40.04 MHz and below 39.96 MHz, but their amplitude is too low to be significant.

either side of the carrier, as shown in Figure 8-12. Because of these multiple sidebands, FM or PM signals inherently occupy a greater bandwidth. The additional sidebands depend on the relationship between the modulating frequency and the frequency deviation. The ratio between the peak carrier frequency deviation and the audio modulating frequency is called the *modulation index.*

Given a constant input level to the modulator, in phase modulation, the modulation index is constant regardless of the modulating frequency. For an FM signal, the modulation index varies with the modulating frequency. In an FM system, the ratio of the maximum carrier-frequency deviation to the highest modulating frequency used is called the *deviation ratio.* Whereas modulation index is a variable that depends on a set of operating conditions, deviation ratio is a constant. The deviation ratio for narrow-band FM is 5000 Hz (maximum deviation) divided by 3000 Hz (maximum modulation frequency) or 1.67.

Frequency Multiplication and Amplification

Since there is no change in the amplitude with modulation, an FM or PM signal can be amplified with a nonlinear amplifier without distortion. This has the advantage that modulation can be introduced in a low-level stage and the signal can then be amplified efficiently.

It is not usually practical to have an oscillator operating in the VHF range or higher, because such an oscillator would not be as stable as one operating at lower frequencies. Common practice is to use an oscillator operating at some submultiple of the desired operating frequency, and then to use frequency multiplication to obtain the proper output. When the modulation is applied to the oscillator, care must be taken to set the proper deviation ratio. When the modulated signal is multiplied to the output frequency, the deviation ratio will also be multiplied by the same factor.

For example, if we are using a 12.20-MHz reactance-modulated oscillator in our FM transmitter operating at 146.40 MHz, and we want the maximum frequency deviation to be 5 kHz, what is the maximum deviation of the oscillator frequency? First we need to know how much multiplication the transmitter is using. We find this by dividing the output frequency by the oscillator frequency:

$$\text{Multiplication factor} = \frac{\text{Transmitter output frequency}}{\text{Oscillator frequency}} \qquad \text{(Equation 8-4)}$$

$$\text{Multiplication factor} = \frac{146.40 \text{ MHz}}{12.20 \text{ MHz}} = 12$$

This means we are multiplying the oscillator frequency by a factor of 12 to bring it up to the operating frequency. So the deviation in the oscillator will also be multiplied by 12.

$$\text{Oscillator deviation} \times \text{Multiplication factor} = \text{Transmitter deviation} \quad \text{(Equation 8-5)}$$

$$\text{Oscillator deviation} \times 12 = 5 \text{ kHz}$$

$$\text{Oscillator deviation} = \frac{5 \text{ kHz}}{12} = 417 \text{ Hz}$$

Frequency multiplication offers a means for obtaining any desired amount of frequency deviation, whether or not the modulator is capable of that much deviation. Overdeviation of an FM transmitter causes "splatter"—out-of-channel emissions that can cause interference to adjacent frequencies.

You should understand that a frequency-multiplier stage is an amplifier that produces harmonics of the input signal. By using a filter to select the desired harmonic, the proper output frequency can be obtained. A frequency multiplier is not the same as a mixer stage, which is used to obtain the desired output frequency in an SSB or CW rig. A mixer simply shifts, or translates, an oscillator frequency to some new frequency.

Bandwidth in FM and PM

With frequency deviation defined as the instantaneous change in frequency for a given signal, we can now define the total bandwidth of the signal. Since the frequency actually swings just as far in both directions, the total frequency swing is equal to twice the deviation. In addition, there are sidebands that increase the bandwidth still further. A good estimate of the bandwidth is twice the maximum frequency deviation plus the maximum modulating audio frequency:

$$\text{Bw} = 2 \times (D + M) \qquad \text{(Equation 8-6)}$$

where
  Bw = bandwidth
  D = maximum frequency deviation
  M = maximum modulating audio frequency

With a transmitter using 5 kHz deviation and a maximum audio frequency of 3 kHz, the total bandwidth is approximately 16 kHz. The actual bandwidth is somewhat greater than this, but it is a good approximation.

[If you are studying for an Element 3A (Technician) exam, turn to Chapter 10 and study questions 3H-6.1, 3H-6.2, 3H-7.1, 3H-12.1, 3H-14.1 and 3H-14.2.

If you are studying for an Element 3B (General) exam, study these questions in Chapter 11: 3H-7.2, 3H-10.1, 3H-10.2, 3H-15.1 and 3H-15.2.]

RADIOTELEPRINTING

Three types of emissions are commonly used by amateurs for transmitting digital codes for automatic reception: A2B, F1B and F2B. Instead of switching a carrier on and off to send information as in A1A, the carrier can be left on continuously and switched between two different frequencies. This is called F1B emission or *frequency-shift keying (FSK)*. F1B is the only one of these emissions authorized for radioteleprinter use below 50 MHz. FSK has the advantage over on-off keying that it gives definite signals for both the on and off states used for digital communications. The on and off states are referred to as "mark" and "space," respectively.

FSK can be produced in two ways. The first method is to change the oscillator frequency back and forth between two frequencies so the carrier also shifts between two frequencies. The second method makes use of the fact that there is no output from an SSB transmitter when there is no modulation. Only when an audio tone is applied to the microphone input is there a corresponding RF signal emitted from the transmitter. By switching between two fixed audio tones, the RF signal is shifted between the mark and space frequencies. The difference in the two RF signals is equal to the difference in frequency of the audio tones.

When using F1B emissions below 50 MHz, no more than 1000 Hz is permitted between the mark and space signals. Above 50 MHz, the frequency shift, in hertz, must not exceed the sending speed, in bauds, of the transmission, or 1000 Hz, whichever is greater.

As the sending speed of an RTTY transmission is increased, the frequency shift between the mark and space tones must also be increased. The electronic device that goes between the RTTY terminal (keyboard and display unit) and the transmitter is called a modem, short for modulator/demodulator. As the sending speed is increased, it becomes more difficult for the modem to distinguish between the two frequencies used to represent mark and space. Increasing the shift between mark and space helps to solve this problem. This also increases the bandwidth of the signal, so the FCC bandwidth restrictions impose practical limits on the keying speed of an RTTY signal. At higher frequencies, where wider bandwidths are permitted, higher keying speeds can be used. Below 28 MHz, the sending speed may not exceed 300 bauds. Between 28 and 50 MHz you may use speeds up to 1200 bauds, between 50 and 220 MHz the speed limit is 19,600 bauds (19.6 kilobauds) and above 220 MHz you may use speeds up to 56 kilobauds. A baud is the unit used to describe transmission speeds for digital signals. For a single-channel transmission, a baud is equivalent to one digital bit of information transmitted per second. So a 300-baud signaling rate represents a transmission rate of 300 bits of digital information per second in a single-channel transmission.

If audio tones are fed into the microphone input of an amplitude-modulated, double-sideband, full-carrier transmitter, A2B signals will be produced. When audio tones are fed to the input of an FM transmitter, F2B emissions are produced. These emissions are called *audio-frequency shift keying (AFSK)*. F2B and A2B emissions are permitted only above 50.1 MHz.

The audio frequencies used for AFSK have been standardized in the US at 2125 Hz and 2295 Hz for 170-Hz shift and 2125 Hz and 2975 Hz for 850-Hz shift. Most amateur operation uses 170-Hz shift. These tones were selected because they are close to the upper limit of the audio passband of most transmitters. This is important because harmonics produced in the tone generator will fall outside the audio passband and, therefore, will be attenuated in the transmitter. When lower-frequency audio tones are used, care must be taken to ensure that harmonics are eliminated from the signal produced by the tone generator. If the audio input to the transmitter contains harmonics, the RF output from the transmitter may contain signals at frequencies other than the desired fundamental. These extra signals can cause interference to other amateur stations, as well as RFI and TVI.

Digital Codes

For many years the only digital code designed for automatic reception that amateurs were allowed to use was the *International Telegraphic Alphabet Number 2 (ITA2),* commonly called *Baudot* or the Murray code. In recent years, two additional codes have been approved for use in the amateur bands. These are the *American National Standard Code for Information Interchange (ASCII)* and *Amateur Teleprinting Over Radio (AMTOR).*

ITA2, or Baudot

ITA2 is a five-bit code, and there are only two distinct conditions for each bit (1 or 0, on or off, mark or space). Therefore, a total of $2 \times 2 \times 2 \times 2 \times 2 = 2^5 = 32$ different code combinations are possible. Because it is necessary to provide 26 Latin letters, 10 numerals and punctuation, the 32 code combinations are not sufficient. This problem is solved by using the codes twice; once in the LETTERS (LTRS) case and once in the FIGURES (FIGS) case. Two special characters, LTRS and FIGS, are used to signal whether characters that follow will be in letters or figures case. The receiving teleprinter remains in the last received case until it receives the signal to change. Control functions such as LTRS, FIGS, carriage return, line feed, space and blank are assigned to both figures and letters case so they may be used in either case. The remaining 26 code combinations have a different meaning, depending on whether they are preceded by a LTRS or FIGS character.

Besides the five data bits, each character also includes a start and a stop pulse. The bits of the Baudot letter D shown in Figure 8-13 are arranged in a left-to-right order, as they would be observed on an oscilloscope.

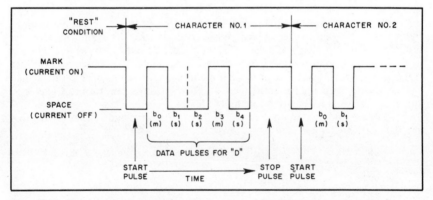

Figure 8-13—Time sequence of the data bits in a typical Baudot character, the letter D.

AMTOR

The AMTOR code is similar to Baudot. In most cases the AMTOR code is derived from Baudot by adding "ones" and "zeros" to give a seven-bit code that consists of four ON or "one" bits and three OFF or "zero" bits. The four/three ratio allows 35 code combinations, which provides three control characters necessary for operation. AMTOR uses these unique characters so the receiving station and the transmitting station can talk to each other to verify the "correctness" of the data as it is being sent. The correspondence between the Baudot code and the AMTOR code can be seen in Table 8-2.

Table 8-2

Conversion Between AMTOR Code and Baudot Code

The codes are transmitted left to right. The higher frequency of the FSK signal is represented by "1."

Baudot Code	Ltrs	Figs	AMTOR Code	Baudot Code	Ltrs	Figs	AMTOR Code
11000	A	—	1110001	00001	T	5	0010111
10011	B	?	0100111	11100	U	7	0111001
01110	C	:	1011100	01111	V	=	0011110
10010	D		1100101	11001	W	2	1110010
10000	E	3	0110101	10111	X	/	0101110
10110	F		1101100	10101	Y	6	1101010
01011	G		1010110	10001	Z	+	1100011
00101	H		1001011	00010	carriage return		0001111
01100	I	8	1011001	01000	line feed		0011011
11010	J	bell	1110100	11111	letters		0101101
11110	K	(	0111100	11011	figures		0110110
01001	L	)	1010011	00100	space		0011101
00111	M	.	1001110	00000	blank		0101011
00110	N	,	1001101		RQ		0110011
00011	O	9	1000111		beta		1100110
01101	P	0	1011010		alpha		1111000
11101	Q	1	0111010		control 1		1010011
01010	R	4	1010101		control 2		0101011
10100	S	'	1101001		control 3		1011001

ASCII

ASCII uses seven data bits and can thus define many more characters. In addition to the seven data bits, an eighth bit can be added to detect errors in a transmitted character. This is called the parity bit. Parity is a property of the data, chosen by the user, to be either even or odd. When used for error detection, the parity bit is made one or zero such that, when all eight bits are examined for the total number of "one" bits, the total number is even or odd, depending on the chosen parity. When transmitting ASCII codes one or more stop bits can also be included in the format.

[If you are studying for an Element 3A (Technician) exam, turn to Chapter 10 and study quesitons 3H-16.1 through 3H-16.3.

If you are studying for an Element 3B (General) exam, study quesitons 3H-16.4 and 3H-16.5 in Chapter 11.]

Key Words

Balanced line—feed line with two conductors having equal but opposite voltages, with neither conductor at ground potential

Balun—a transformer used between a BALanced and an UNbalanced system, such as for feeding a balanced antenna with an unbalanced feed line

Coaxial cable—feed line with a central conductor surrounded by plastic, foam or gaseous insulation, which in turn is covered by a shielding conductor and the entire cable is covered with vinyl insulation

Cubical quad antenna—an antenna built with its elements in the shape of four-sided loops

Delta loop antenna—a variation of the cubical quad antenna with triangular elements

Director—a parasitic element in "front" of the driven element in a multielement antenna

Driven element—the element connected directly to the feed line in a multielement antenna

Feed line—the wire or cable used to connect an antenna to the transmitter and receiver

Front-to-back ratio—the energy radiated from the front of a directive antenna divided by the energy radiated from the back of the antenna

Gain—an increase in the effective power radiated by an antenna in a certain desired direction, or an increase in received signal strength from a certain direction. This is at the expense of power radiated in, or signal strength received from, other directions.

Gamma match—a method of matching coaxial feed line to the driven element of a multielement array

Ground-plane antenna—a vertical antenna built with a central radiating element one-quarter-wavelength long and several radials extending horizontally from the base. The radials are slightly longer than one-quarter wave, and may droop toward the ground.

Half-wavelength dipole antenna—a fundamental antenna one-half wavelength long at the desired operating frequency, and connected to the feed line at the center. This is a popular amateur antenna.

Horizontally polarized wave—an electromagnetic wave with its electric lines of force parallel to the ground

Major lobe—the direction of maximum radiated field strength from an antenna

Parallel-conductor feed line—feed line constructed of two wires held a constant distance apart; either encased in plastic or constructed with insulating spacers placed at intervals along the line

Parasitic element—part of a directive antenna that derives energy from mutual coupling with the driven element. Parasitic elements are not connected directly to the feed line.

Polarization—the orientation of the electric lines of force in a radio wave, with respect to the surface of the earth

Quarter-wavelength vertical antenna—an antenna constructed of a quarter-wavelength-long radiating element placed perpendicular to the earth. (See Ground-plane antenna.)

Reflector—a parasitic element placed "behind" the driven element in a directive antenna

Standing wave ratio—the ratio of maximum voltage to minimum voltage along a feed line. Also the ratio of antenna impedance to feed line impedance when the antenna is a purely resistive load.

Transmission line—see Feed line

Twin lead—see Parallel-conductor feed line

Unbalanced line—feed line with one conductor at ground potential, such as coaxial cable.

Velocity factor—the ratio of the velocity of a radio wave in a conductor to the velocity of the wave in a vacuum

Vertically polarized wave—a radio wave that has its electric lines of force perpendicular to the surface of the earth

Yagi antenna—a directive antenna made with a half-wavelength driven element, and two or more parasitic elements arranged in the same horizontal plane

Chapter 9

Antennas and Transmission Lines

A n antenna system can be considered to include the antenna proper (what is actually radiating the signal), the feed line, and any coupling devices or matching networks used for transferring power from the transmitter to the line and from the line to the antenna. In some simple systems the feed line may be part of the antenna, and coupling devices may not be used. An applicant for the Technician or General class license should know the basics of transmission lines and the commonly used Amateur Radio antennas.

WAVE CHARACTERISTICS AND POLARIZATION

An electromagnetic wave consists of moving electric and magnetic fields. Remember that a field is an invisible force of nature. We can't see radio waves—the best we can do is to show a representation of where the energy is in the electric and magnetic fields. We did this in Chapter 5 to show the magnetic flux around a coil, and the electric field in a capacitor. We can visualize a travelling radio wave as looking something like Figure 9-1. The lines of electric and magnetic force are at right angles to each other, and are also perpendicular to the direction of travel. These fields can have any position with respect to the earth.

Polarization is defined by the direction of the electric lines of force in a radio wave. If the electric lines of force are parallel to the surface of the earth, the wave

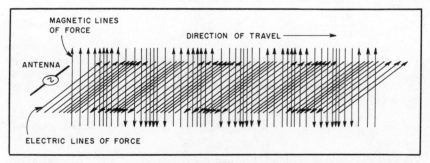

Figure 9-1—A horizontally polarized electromagnetic wave. The electrical lines of force are parallel to the ground. The fields are stronger where the arrows are closer together. The points of maximum field strength correspond to the points of maximum amplitude for the voltage producing the RF signal.

is called a *horizontally polarized wave*. Similarly, if the electric lines of force are perpendicular to the earth, the wave is called a *vertically polarized wave*.

The polarization of a radio wave as it leaves the antenna is determined by the orientation of the antenna. For example, a half-wavelength dipole parallel to the surface of the earth transmits a wave that is horizontally polarized. An amateur mobile whip antenna, mounted vertically on an automobile, transmits a wave that is vertically polarized.

After radio waves have travelled some distance from the transmitting antenna and have interacted with their surroundings and with the ionosphere, the polarization may change. Longer wavelengths will typically retain their polarization, while shorter wavelengths will often change polarization as they interact with the ionosphere. These changes can occur quite rapidly. Most man-made noise tends to be vertically polarized; thus, a horizontally polarized antenna will receive less noise of this type than a vertical antenna will.

THE HALF-WAVELENGTH DIPOLE ANTENNA

A popular amateur antenna is a wire whose total length is approximately equal to half the wavelength of the transmitted (or received) signal. This antenna is so basic, in fact, that it is the unit from which many more complex forms of antennas are constructed. This basic antenna is called the *half-wavelength dipole antenna*.

The wavelength of a radio wave is related to the speed of light and the frequency of the wave by the equation:

$$\lambda = \frac{c}{f} \qquad \text{(Equation 9-1)}$$

where
λ = one wavelength in meters
c = the speed of light in meters per second
f = the frequency in hertz

The speed of light in a vacuum is 3×10^8 meters per second, and if we use the frequency in megahertz (hertz $\times 10^6$) the equation reduces to:

$$\lambda = \frac{3 \times 10^2}{f \text{ (megahertz)}} = \frac{300}{f \text{ (MHz)}} \qquad \text{(Equation 9-2)}$$

So a half wavelength in meters is given by the formula:

$$\lambda/2 = \frac{150}{f \text{ (megahertz)}} \qquad \text{(Equation 9-3)}$$

The length of a dipole antenna will be somewhat less than this because radio waves move more slowly in metal than in a vacuum. If we convert from metric to US Customary units we can derive an approximate relation between frequency (f) and length (in feet) for a wire antenna:

$$L(\text{ft}) = \frac{468}{f \text{ (MHz)}} \qquad \text{(Equation 9-4)}$$

For example, using Equation 9-2 to find the wavelength of a 7.15-MHz signal:

$$\lambda = \frac{300}{f \text{ (MHz)}} = \frac{300}{7.15 \text{ MHz}} = 42.0 \text{ meters}$$

Then, using Equation 9-4 to find the length of a half-wavelength dipole for that frequency:

$$L(ft) = \frac{468}{f\ (MHz)} = \frac{468}{7.15\ MHz} = 65.45\ ft$$

The diameter of the radiating element also affects the frequency at which the antenna will resonate. Antennas made from tubing will be shorter than antennas made from wire for resonance at the same frequency.

RADIATION CHARACTERISTICS OF ANTENNAS

The radiation from a dipole antenna is not uniform in all directions. It is strongest in directions perpendicular to the wire, and nearly zero along the axis of the wire, as shown in Figure 9-2. This pattern is affected by the antenna height above

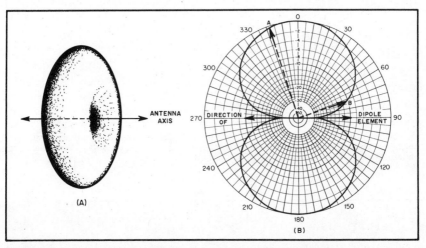

Figure 9-2—At A, the solid (three-dimensional) directive pattern of a dipole. At B, the plane directive diagram of a dipole. The solid line shows the direction of the wire. Note that the pattern is bidirectional; the dipole radiates equally well in two directions.

ground. The half-wavelength dipole exhibits excellent directivity when placed at least one-half wavelength above the ground, as shown by the pattern in Figure 9-3A. A dipole cut for the 40-meter band, for example, must be mounted at least 20 meters above ground to exhibit such directivity. If the antenna is lower, the directivity effects are lessened. Figure 9-3B shows the radiation pattern for a dipole mounted only one-quarter wavelength above the ground. Notice that the pattern is more nearly circular.

Despite this apparent disadvantage, the dipole is the most common and practical antenna for the 80- and 40-meter bands. The impedance at the feed point of a half-wavelength dipole suspended horizontally at least one-quarter wavelength above ground is close to 70 ohms, so a dipole can be fed directly with 75-ohm coaxial cable. The feed point impedance will be less than 70 ohms if the antenna is mounted less than one-quarter wavelength above the ground, as shown in Figure 9-4. Coaxial cable with a characteristic impedance of 50 ohms is often used as the feed line to dipole antennas.

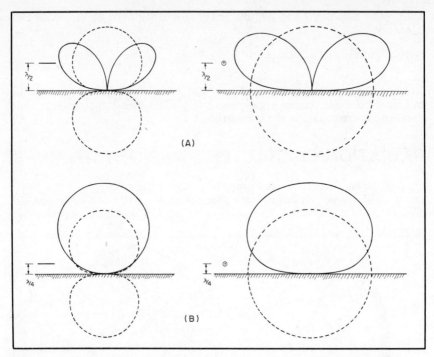

Figure 9-3—Effect of ground on the radiation from a horizontal half-wavelength antenna, for heights of one-half (A) and one-quarter (B) wavelength. The dashed lines show what the pattern would be if there were no reflection from the ground.

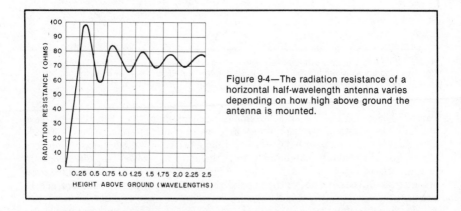

Figure 9-4—The radiation resistance of a horizontal half-wavelength antenna varies depending on how high above ground the antenna is mounted.

One variation of the half-wavelength dipole is the inverted V antenna, shown in Figure 9-5. In this antenna, the ends of the dipole are lowered with respect to its center. This type of antenna works best when the angle between the two antenna wires is 90° or more.

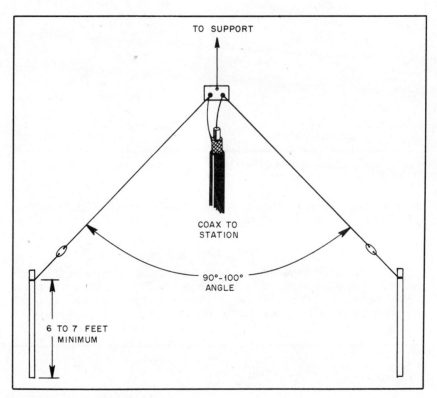

Figure 9-5—The inverted-V dipole antenna.

MULTIPLE-ELEMENT ARRAYS

The radiation pattern of the half-wavelength dipole antenna is described as bidirectional—it radiates equally well in two directions. A unidirectional pattern (maximum radiation in one direction) can be produced by coupling the half-wavelength antenna to additional elements. The additional elements may be made of wire or metal tubing.

Parasitic Excitation

In most multiple-element antennas, the additional elements are not directly connected to the feed line. They receive power by mutual coupling from the *driven element* (the element connected to the feed line). They then reradiate it in the proper phase relationship to achieve gain or directivity over a simple half-wavelength dipole. These elements are called *parasitic elements*.

There are two types of parasitic elements. A *director* is generally shorter than the driven element and is located at the "front" of the antenna. A *reflector* is generally longer than the driven element and is located at the "back" of the antenna. See Figure 9-6. The direction of maximum radiation from a parasitic antenna travels from the reflector through the driven element and the director. The term *major lobe* refers to the region of maximum radiation from a directional antenna. The major lobe is

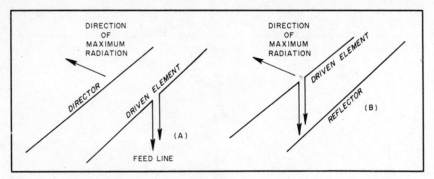

Figure 9-6—In a directive antenna, the reflector element is placed "behind" the driven element. The director goes in "front" of the driven element.

also sometimes referred to as the main lobe. If communication in different directions is desired, provision must be made to rotate the array in the azimuth, or horizontal, plane.

Yagi Arrays

The *Yagi* arrays in Figure 9-7 are examples of antennas that make use of parasitic elements to produce a unidirectional radiation pattern. Though typical HF antennas of this type have three elements, some may have six or more. Multiband Yagi antennas have many elements, some of which work on some frequencies and others that

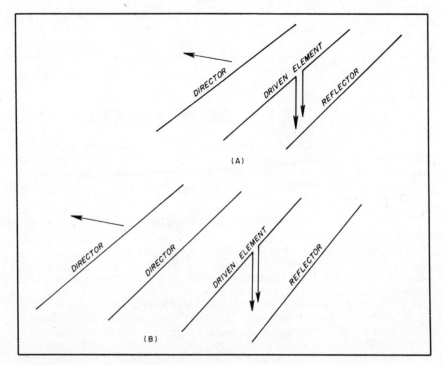

Figure 9-7—At A, a three-element Yagi. At B, a four-element Yagi with two directors.

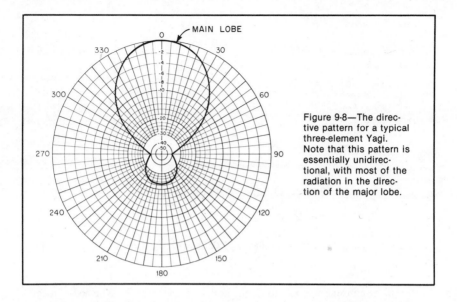

Figure 9-8—The directive pattern for a typical three-element Yagi. Note that this pattern is essentially unidirectional, with most of the radiation in the direction of the major lobe.

are used for different frequencies. The radiation pattern of the Yagi antenna is shown in Figure 9-8. This pattern indicates that the antenna will reject signals coming from the sides and back, selecting mainly those signals from a desired direction. There are several types of Yagi antennas, and many different methods of connecting the feed line to the driven element. The most common feed system, shown in Figure 9-9, is called the *gamma match*.

The length of the driven element in the most common type of Yagi antenna is approximately one-half wavelength. This means that Yagi antennas are most often used for the 20-meter band and higher frequencies. Below 20 meters, the physical dimensions become so large that the antennas require special construction techniques and heavy-duty supporting towers and rotators. Yagis for 40 meters are fairly common, but 80-meter rotatable

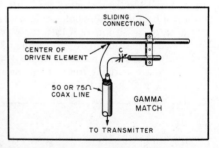

Figure 9-9—The gamma match—the matching system best suited to matching Yagi antennas to unbalanced (coaxial) feed lines.

Yagis are few and far between. To overcome the mechanical difficulties, some amateurs build nonrotatable Yagi antennas for 40 and 80 meters with elements made of wire supported on both ends.

In a three-element beam, the director will be approximately 5% (0.05) shorter than one-half wavelength, and the reflector will be approximately 5% longer than one-half wavelength. We can derive a set of equations to calculate these element lengths.

To find the length of a director or reflector element for a Yagi with a 14.06-MHz center frequency, we can use Equations 9-4, 9-5 and 9-6:

$$L_{director} = L_{driven} (1 - 0.05)$$ (Equation 9-5)

or

$$L_{director} = L_{driven} \times 0.95$$

$$L_{reflector} = L_{driven} (1 + 0.05) \qquad \text{(Equation 9-6)}$$

or

$$L_{reflector} = L_{driven} \times 1.05$$

$$L \text{ (ft)} = \frac{468}{f \text{ (MHz)}} \qquad \text{(Equation 9-4)}$$

$$L_{driven} = \frac{468}{14.06 \text{ MHz}} = 33.3 \text{ ft}$$

$$L_{director} = L_{driven} \times 0.95 = 33.3 \text{ ft} \times 0.95$$

$$L_{director} = 31.6 \text{ ft}$$

$$L_{reflector} = L_{driven} \times 1.05 = 33.3 \text{ ft} \times 1.05$$

$$L_{reflector} = 34.97 \text{ ft}$$

There can be considerable variation on these lengths, however. The actual lengths depend on the spacing between elements, the diameter of the elements, and whether the elements are made from tapered or cylindrical tubing.

The polarization of the signal transmitted from a Yagi antenna is determined by the placement of the antenna relative to the surface of the earth. With the elements parallel to the surface of the earth, as shown in Figure 9-10A, the transmitted wave will be horizontally polarized. With the elements perpendicular to the surface of the earth, as in Figure 9-10B, the wave will have vertical polarization.

[If you are studying for an Element 3A (Technician) exam, turn to Chapter 10 and study questions 3I-1.1, 3I-1.2, 3I-1.4, 3I-1.5, 3I-1.6 and 3I-1.8.

If you are studying for an Element 3B (General) exam, study questions 3I-1.3 and 3I-1.7 in Chapter 11.]

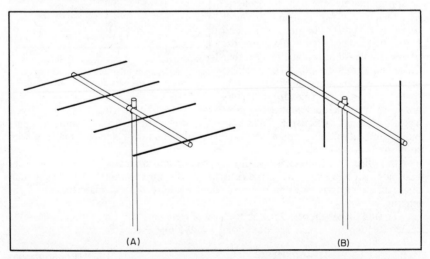

(A)   (B)

Figure 9-10—The orientation of the elements in a Yagi antenna determines the polarization of the transmitted wave. If the elements are horizontal as shown at A, the wave will have horizontal polarization. Vertical mounting, shown at B, produces a vertically polarized wave.

Cubical Quad Antennas

A second type of antenna that uses a parasitic element is called a *cubical quad* antenna or simply a *quad*. The elements of the quad antenna are full-wavelength loops. A typical quad, shown in Figure 9-11, has two elements—a driven element and a reflector. A two-element quad could also use a driven element with a director. More elements, such as a reflector and one or more directors, can be added to increase the directivity of the antenna. By installing additional loops of the right dimensions, we can work different frequency bands with the same antenna. This is commonly done as a modification to a 20-meter quad to provide 15- and 10-meter coverage. The elements of the quad are usually square, with each side a quarter wavelength long. The total lengths of the elements are calculated as follows:

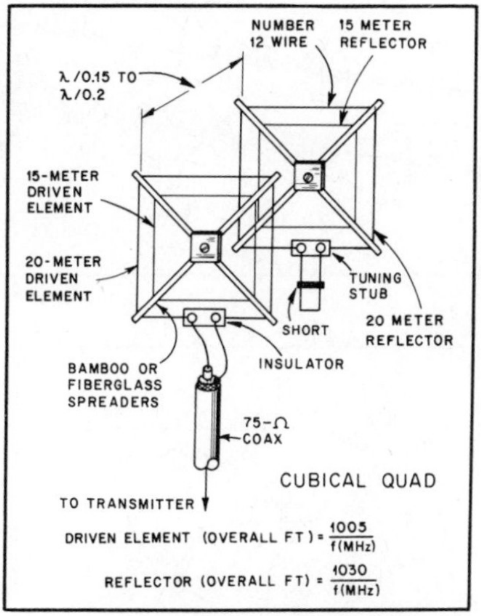

Circumference of driven element:

$$L(ft) = \frac{1005}{f\ (MHz)}$$

(Equation 9-7)

Circumference of director element:

$$L(ft) = \frac{975}{f\ (MHz)}$$

(Equation 9-8)

Circumference of reflector element:

$$L(ft) = \frac{1030}{f\ (MHz)}$$

(Equation 9-9)

Figure 9-11—The cubical quad antenna. The total length of the driven element can be found using Equation 9-7, and the antenna can be fed directly with coaxial cable.

So for a 21.25-MHz cubical quad, the element lengths would be:

$$\text{driven element} = \frac{1005}{21.25} = 47.3\ ft$$

$$\text{director} = \frac{975}{21.25} = 45.9\ ft$$

$$\text{reflector} = \frac{1030}{21.25} = 48.5\ ft$$

Remember that these equations give the total length of the elements. To find the length of each side of the antenna, we must divide the total length by 4.

The polarization of the signal from a quad antenna is determined by where the feed point is located on the driven element, as shown in Figure 9-12. With the feed-point located in the center of a side parallel to the earth's surface, the transmitted wave will be horizontally polarized; fed in the center of a perpendicular side, the transmitted wave will be vertically polarized. If we turn the antenna 45°, so it looks

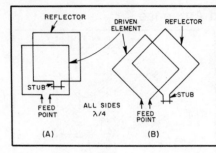

Figure 9-12—The feed point of a quad antenna determines the polarization. Fed in the middle of the bottom side (as shown at A) or at the bottom corner (as shown at B), the antenna produces a horizontally polarized wave. Vertical polarization is produced by feeding the antenna at the side in either case.

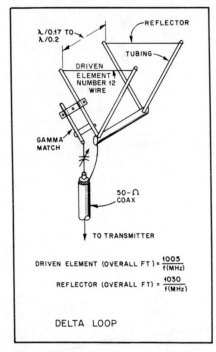

DRIVEN ELEMENT (OVERALL FT) = $\dfrac{1005}{f(MHz)}$

REFLECTOR (OVERALL FT) = $\dfrac{1030}{f(MHz)}$

DELTA LOOP

Figure 9-13—A delta loop antenna. The total length of the driven element can be found using the same equation as used for a quad antenna. This antenna is fed with a gamma match.

like a diamond, and feed it at the bottom corner, the transmitted wave will be horizontally polarized. If the antenna is fed at a side corner, the transmitted wave will be vertically polarized.

Polarization is especially important when building antennas for use at VHF and UHF. Propagation at these frequencies is generally line of sight, and the polarization of a signal does not change from transmitting antenna to receiving antenna. Best signal reception will occur when both transmitting and receiving stations use the same polarization. On the HF bands, the polarization of the signal may change many times as it passes through the ionosphere, so antenna polarization is not as important.

A variation on the quad is the delta loop antenna, shown in Figure 9-13. In a delta loop, the elements are triangular rather than square. The same equations used for the quad will work to calculate the element lengths for the delta loop. To find the length of each side of the delta loop antenna, the total length is divided by 3, because of the triangular shape of the elements. The radiation pattern of a typical quad is similar to that of the Yagi shown in Figure 9-8.

Front-to-Back Ratio

Front-to-back ratio is the ratio between the power radiated in the direction of maximum radiation (the "front" of the antenna) and the power radiated in the reverse direction (the "back" of the antenna). If you point your antenna directly at a receiving station and transmit, and then turn the antenna 180° and transmit again, the difference in the received signal strength is the front-to-back ratio, which is usually expressed in decibels. An increase in the front-to-back ratio is brought about by adjusting the length of the parasitic elements or their position relative to the driven element and each other. The tuning condition that gives maximum attenuation to the rear is considerably more critical than the condition for maximum forward gain.

Gain versus Spacing

When we speak of the *gain* of an antenna, we mean gain referenced to another antenna, used for comparison. One common reference antenna is an ideal half-wavelength dipole in free space. The gain of an antenna with parasitic elements varies with the boom length of the antenna, the spacing between elements, and the length and the number of parasitic elements. The maximum front-to-back ratio seldom, if ever, occurs at the same conditions that yield maximum forward gain, and it is frequently necessary to sacrifice some forward gain to get the greatest front-to-back ratio, or vice versa. Changing the tuning and spacing of the elements also affects the impedance of the driven element and the bandwidth of the antenna. The most effective method to increase the bandwidth of a parasitic beam antenna is to use larger-diameter elements.

Theoretically, an optimized three-element (director, driven element and reflector) Yagi antenna has a gain of slightly more than 7 dB over a half-wavelength dipole in free space. A typical two-element quad has a gain of about 6.5 dB over a half-wavelength dipole.

[If you are studying for an Element 3A (Technician) exam, turn to Chapter 10 and study the following questions: 3I-2.2 through 3I-2.6, 3I-4.1 through 3I-4.8 and question 3I-6.7.

If you are studying for an Element 3B (General) exam, study these questions in Chapter 11: 3I-1.9, 3I-2.1, 3I-3.1 through 3I-3.9 and 3I-6.1 through 3I-6.6.]

VERTICAL ANTENNAS

Another popular antenna is the *quarter-wavelength vertical antenna*. It is often used to obtain low-angle radiation when a beam or dipole cannot be placed far enough above ground. Low-angle radiation refers to signals that travel closer to the horizon, rather than signals that are high above the horizon. Low-angle radiation is usually advantageous when you are attempting to contact distant stations.

You might also want to use a vertical antenna if you don't have two supports separated by an appropriate distance to hold the ends of a dipole antenna. As shown in Figure 9-14, the quarter-wavelength vertical antenna is half of a half-wavelength antenna mounted vertically. Conductivity in the ground produces an "image" of the missing quarter wavelength. An excellent ground connection is necessary for the most effective operation of a vertical antenna.

In most installations, ground conductivity is inadequate, so an artificial ground must be made from wires placed along the ground near the base of the antenna, as shown in Figure 9-14C. These wires, called radials, are usually one-quarter wavelength long or longer. Depending on ground conductivity, 8, 16, 32 or more radials may be required to form an effective ground. For best results, the vertical radiator should be located away from nearby conductive objects.

Vertical antennas can be combined to form arrays in which each element is fed with a signal from the transmitter. By controlling the phase relationships of the signals fed to each element, a variety of radiation patterns can be obtained.

The radiation resistance of a quarter-wavelength antenna over a perfect ground is about 36 ohms. The feed-point impedance of a real quarter-wavelength antenna may vary between 30 and 100 ohms, depending on ground resistance. The length of the antenna is given by the formula:

$$L \text{ (ft)} = \frac{234}{f \text{ (MHz)}} \qquad \text{(Equation 9-10)}$$

Let's try calculating the length of a quarter-wavelength vertical antenna for use on 28.6 MHz:

$$L \text{ (ft)} = \frac{234}{28.6 \text{ MHz}} = 8.18 \text{ ft}$$

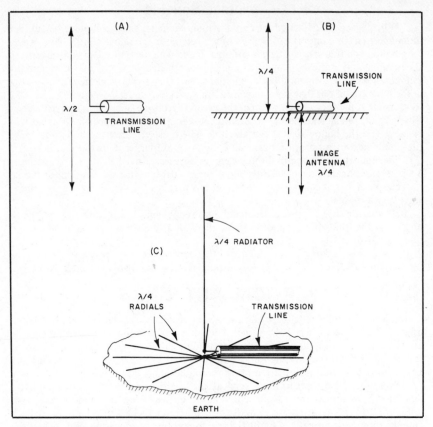

Figure 9-14—A half-wavelength vertical antenna is shown at A. At B is its grounded quarter-wavelength counterpart. The missing quarter wavelength can be considered to be supplied by the image in ground with good conductivity. A practical quarter-wave vertical antenna installation will require the use of several ground radials, as shown at C.

When tubing is used for the antenna, or when guy wires are used to reinforce the structure, the length given by Equation 9-10 is likely to be long by a few percent. As with any antenna, use the formula as a starting length for constructing your version, and then adjust it as needed to make the antenna resonant at the desired operating frequency.

The Ground-Plane Antenna

The *ground-plane antenna* is a vertical antenna that uses an artificial metallic ground consisting of metal rods or wires either perpendicular to, or sloping from, the antenna base and extending outward, as shown in Figure 9-15. The entire antenna, including the radials, is mounted above the ground. This permits the antenna to be mounted at a height clear of surrounding obstructions such as trees and buildings. The ground plane also requires only 3 or 4 radials to operate properly, as opposed to the large number required for some ground-mounted installations.

With the radials perpendicular to the antenna, the feed-point impedance is around 35 ohms. Sloping the radials downward raises the feed-point impedance of the antenna somewhat. If the radials are angled down at about a 45° angle, the feed-point im-

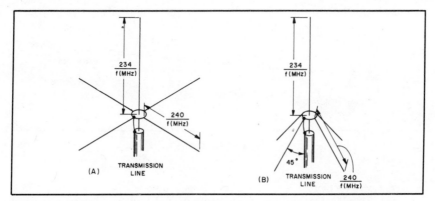

Figure 9-15—The ground-plane antenna. Drooping the radials (as shown at B) raises the feed-point impedance closer to 50 ohms.

pedance is around 50 ohms, which makes it easier to match the antenna to available transmission lines. Unlike other quarter-wavelength vertical antennas without an artificial ground, the ground-plane antenna will give low-angle radiation regardless of the height above actual ground. The radials should be slightly longer than the radiating element; the formula for the length of the radials is:

$$L \text{ (ft)} = \frac{240}{f \text{ (MHz)}}$$ (Equation 9-11)

We can use this equation to calculate the length of the radials needed for our 28.6-MHz vertical antenna in the previous example:

$$L \text{ (ft)} = \frac{240}{28.6 \text{ (MHz)}} = 8.39 \text{ ft}$$

[If you are studying for an Element 3B (General) exam, turn to Chapter 11 and study the questions in subelement 3I that begin 3I-5.]

Transmission Lines

A *transmission line* (or *feed line*) is used to transfer power from the transmitter to the antenna. The two basic types of transmission lines generally used, *parallel-conductor feed line* and *coaxial cable,* can be constructed in a variety of forms. Both types can be divided into two classes: those in which the majority of the space between the conductors is air, and those in which the conductors are embedded in and separated by a solid plastic or foam insulation (dielectric).

Air-insulated transmission lines have the lowest loss per unit length (usually expressed in dB/100 ft). Adding a solid dielectric between the conductors increases the losses in the feed line, and decreases the maximum voltage the feed line can withstand before arcing. In general, as frequency increases, feed-line losses become greater.

A typical type of construction used for parallel-conductor or "ladder line" air-insulated transmission lines is shown in Figure 9-16. The two wires are supported a fixed distance apart by means of insulating rods called spacers. Spacers are commonly made from phenolic, polystyrene, isolantite or Lucite™. The spacers generally vary in length from 1 to 6 inches. The shorter lengths are desirable at the higher frequencies so that the conductors are held a small fraction of a wavelength apart and radiation from the transmission line will be minimized. Spacers are placed along the line at intervals that are small enough to prevent the two lines from moving appreciably

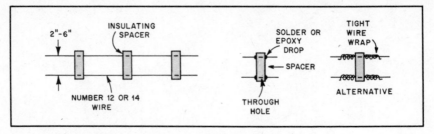

Figure 9-16—Typical open-wire line construction is shown.

with respect to each other. This type of line is sometimes referred to as "open-wire" feed line. An advantage of this type of transmission line is that it can be operated at a high SWR and still retain its low-loss properties.

The characteristic impedance of an air-insulated parallel-conductor line depends on the diameter of the wires used in the feed line and the spacing between them. The greater the spacing between the conductors the higher the characteristic impedance of the feed line. The impedance decreases, however, as the size of the conductors increases. The characteristic impedance of a feed line is not affected by the length of the line.

Solid-Dielectric Feed Lines

Transmission lines in which the conductors are separated by a flexible dielectric have several advantages over air-dielectric line: They are less bulky, maintain more uniform spacing between conductors, are generally easier to install and are neater in appearance. Both parallel-conductor and coaxial lines are available with this type of insulation.

One disadvantage of these types of lines is that the power loss per unit length is greater than air-insulated lines because of the dielectric. As the frequency increases, the dielectric losses become greater. The power loss causes heating of the dielectric. If the heating is great enough—as may be the case with high power or a high SWR—the dielectric may actually melt, or arcing may occur inside the line.

One common parallel-conductor feed line has a characteristic impedance of 300 ohms. This type of line is used for TV antenna feed line, and is usually called *twin lead*. Twin lead consists of two number 20 wires that are molded into the edges of a polyethylene ribbon about a half-inch wide. The presence of the solid dielectric lowers the characteristic impedance of the line as compared to the same conductors in air. The fact that part of the field between the conductors exists outside the solid dielectric leads to operating disadvantages. Dirt or moisture on the surface of the ribbon tends to change the characteristic impedance. Weather effects can be minimized, however, by coating the feed line with silicone grease or car wax. In any case, the changes in the impedance will not be very serious if the line is terminated in its characteristic impedance (Z_0). If there is a considerable standing-wave ratio, however, then small changes in Z_0 may cause wide fluctuations of the input impedance.

Coaxial Cable

Coaxial cables are available in flexible and semiflexible formats, but the fundamental design is the same for all types. Some coaxial cables have stranded-wire center conductors, while others employ a solid copper conductor. Coaxial cables commonly used by radio amateurs have a characteristic impedance of 50 ohms or 75 ohms. The outer conductor or shield may be a single layer of copper braid, a double layer of braid (more effective shielding), or solid copper or aluminum (most effective

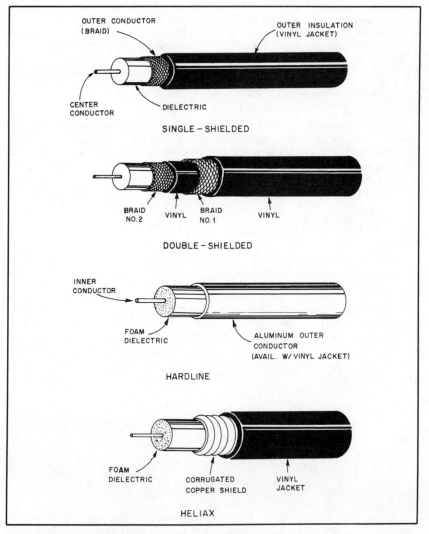

Figure 9-17—Some of the common types of coaxial-cable feed line.

shielding). An outer insulating jacket (usually vinyl) provides protection from dirt, moisture and chemicals. Typical examples of coaxial cables are shown in Figure 9-17. Exposure of the dielectric to moisture and chemicals will cause it to deteriorate over time and increase the electrical losses in the line.

The characteristic impedance of a coaxial line depends on the diameter of the center conductor, the inside diameter of the shield braid and the distance between them. The characteristic impedance of coaxial cables increases for larger-diameter shield braids, but it decreases for larger-diameter center conductors.

The larger the diameter, the higher the power capability of the line because of the increased dielectric thickness and conductor size. In general, losses decrease as the cable diameter increases, because there is less power lost in the conductor. As the frequency increases, the conductor and dielectric losses become greater, causing more attenuation of the signal in the cable. Attenuation is not affected by the characteristic impedance of the line, at least if the spacing between the conductors in the coaxial cable is a small fraction of a wavelength at the operating frequency. If the spacing between conductors is a significant portion of a wavelength, the line will begin to act as a waveguide. At frequencies below about 1.5 GHz this is not a problem.

Electrical Length

Whenever reference is made to a line as being so many wavelengths (such as a "half-wavelength" or "quarter-wavelength") long, it is understood that what is meant is the electrical length of the line. The electrical length of a feed line (expressed in wavelengths) will not be the same as its physical length (expressed in distance units such as feet or meters), and will depend mainly on the type of dielectric used in the feed line. The physical length corresponding to an electrical wavelength is given by:

$$\lambda \text{ (ft)} = \frac{984 \times V}{f \text{ (MHz)}} \qquad \text{(Equation 9-12)}$$

where

f = frequency in megahertz
V = velocity factor of the feed line
λ = wavelength of the radio wave

The *velocity factor* is the ratio of the actual velocity of a radio wave along the line to the velocity in free space. Because quarter-wavelength lines are used frequently, it is convenient to calculate the length of a quarter-wave feed line directly from the equation:

$$\text{length (ft)} = \frac{246 \times V}{f \text{ (MHz)}} \qquad \text{(Equation 9-13)}$$

where f and V are defined as in Equation 9-12.

For example, using coaxial cable with a velocity factor of 0.66, the length of a quarter-wave section at 10.12 MHz would be:

$$\text{length (ft)} = \frac{246 \times 0.66}{f \text{ (MHz)}} = \frac{162.36}{10.12} = 16.04 \text{ ft}$$

[If you are studying for an Element 3A (Technician) exam, turn to Chapter 10 and study these questions: 3I-11.3, 3I-11.5, 3I-11.6, 3I-11.9 and 3I-11.11.

If you are studying for an Element 3B (General) exam, study the following questions in Chapter 11: 3I-7.1 through 3I-7.4, 3I-11.1, 3I-11.2, 3I-11.4, 3I-11.7, 3I-11.8, 3I-11.10 and 3I-11.12.]

Standing-Wave Ratio (SWR)

In a perfect antenna system, all the power put into the feed line would be radiated by the antenna. A practical antenna system will behave most like this ideal when the feed-line impedance is matched to the antenna feed-point impedance. Some power will still be lost in the feed line; this power is converted to heat.

If the antenna feed-point impedance does not exactly match the characteristic impedance of the feed line, some power will be reflected back down the feed line from the antenna. A mismatch is said to exist—the greater the mismatch, the more power will be reflected by the antenna. Some of the reflected power is dissipated as heat. In a lossy line more heat is produced than in a lower-loss line. The reflected power is reflected again by the transmitter and goes back to the antenna, where some power will be radiated, and some reflected back to the transmitter. This process con-

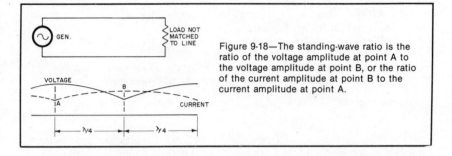

Figure 9-18—The standing-wave ratio is the ratio of the voltage amplitude at point A to the voltage amplitude at point B, or the ratio of the current amplitude at point B to the current amplitude at point A.

tinues until, eventually, all the power is either radiated by the antenna or lost as heat in the feed line.

The reflected power creates what is called a standing wave. If we measure either the voltage or current at several points along the feed line, we find that it varies from a maximum value to a minimum value at intervals of one-quarter wavelength, as shown in Figure 9-18. The standing wave is created when energy reflected from the antenna meets the forward energy from the transmitter—the two waves alternately cancel and reinforce each other.

The *standing-wave ratio* or *SWR* is defined as the ratio of the maximum voltage to the minimum voltage in the standing wave:

$$SWR = \frac{E_{max}}{E_{min}}$$ (Equation 9-14)

An SWR of 1:1 means you have no reflected power. The voltage and current are constant at any point along the feed line, and the line is said to be "flat." If the load is completely resistive, then the SWR can be calculated by dividing the line characteristic impedance by the load resistance, or vice versa—whichever gives a value greater than one:

$$SWR = \frac{Z_0}{R} \text{ or } SWR = \frac{R}{Z_0}$$ (Equation 9-15)

where
Z_0 = characteristic impedance of the transmission line
R = load resistance (not reactance)

For example, if you feed a 100-ohm antenna with 50-ohm transmission line the SWR is 100 / 50 or 2:1. Similarly if the impedance of the antenna is 25 ohms the SWR is 50 / 25 or 2:1.

When a high standing-wave ratio condition exists, losses in the feed line are increased. This is because of the multiple reflections from the antenna and transmitter. Each time the transmitted power has to travel up and down the feed line, a little more energy is lost as heat. This effect is not as great as some people believe, however. If the line loss is less than 2 dB (such as for 100 feet of RG-8 or RG-58 cables up to about 30 MHz), the SWR would have to be greater than 3:1 to add an extra decibel of loss because of the SWR.

When a transmission line is terminated in a resistance equal to its characteristic impedance, maximum power is delivered to the antenna, and transmission line losses are minimized. This would be an ideal condition, although such a perfect match is seldom realized in a practical antenna system.

When you are using a directional wattmeter to measure the output power from your transmitter, remember that the reflected power must be subtracted from the forward reading to give you the true forward power. This is because the power reflected from the antenna will again be reflected by the transmitter, adding to the forward power reading on the meter. For example, if your transmitter power output is 100 watts and 10 watts are reflected from the antenna, those 10 watts will be added into the forward reading on your meter when they are reflected from the transmitter, and your wattmeter reading would be incorrect. To find true forward power you must subtract the reflected measurement from the forward power measurement:

True forward power = Forward power reading − Reflected power reading
(Equation 9-16)

For example, suppose you had a wattmeter connected in the line from your transmitter, and it gave a forward-power reading of 85 watts and a reflected-power reading of 15 watts. What is the true power out of your transmitter?

True forward power = 85 W − 15 W = 70 watts.

Impedance Matching

Most currently manufactured Amateur Radio transmitters are designed to be connected to antenna systems with impedances of 50 to 75 ohms. Some transmitters with vacuum-tube power amplifiers that have a tunable output circuit, however, can be adjusted to match impedances over a greater range. Depending on the type of antenna you are using, you may have to improve the match between the transmission line and the antenna.

A matching network (Transmatch) can be used close to the transmitter to match the 50-ohm output impedance of the transmitter to the impedance of your antenna system. It can be designed to transform the single-ended or *unbalanced line* output (one side grounded to the chassis) to a *balanced line* output for use with a parallel transmission line.

Balanced feed lines may be tuned by means of the Transmatch circuit shown in Figure 9-19. The coil taps and capacitor settings are adjusted for lowest SWR. Higher impedance loads are tapped farther out from the center of the coil. The split-stator capacitor (C1 and C2) is used because it provides two symmetric sections that are tuned simultaneously with a single control.

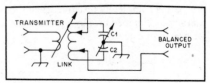

Figure 9-19—An impedance-matching network suitable for use with balanced lines.

When a coaxial cable is used to feed a half-wave dipole, a transformer, called a *balun* (BALanced to UNbalanced), can be used to decouple the transmission line from the antenna. This prevents the feed line from radiating a signal, which could distort the radiation pattern of the antenna.

It is important to note that even though a transmitter may "see" a pure 50-ohm load, and the SWR between the transmitter and matching device is 1:1, the SWR measured in the transmission line between the matching device and the antenna will not be 1:1. Inserting a matching device between the transmission line and the transmitter does not affect the SWR in the transmission line between the matching device and the antenna; the matching device only provides a 50-ohm resistive load for the transmitter. The SWR meter should be placed between the transmitter and the Transmatch to adjust the matching network properly.

[If you are studying for an Element 3A (Technician) exam, turn to Chapter 10 and study questions 3D-5.3 and 3D-5.4. Also study questions 3I-8.1 through 3I-8.3, 3I-9.1, 3I-9.2, questions 3I-10.1 through 3I-10.4, 3I-12.3 and 3I-12.4.

If you are studying for an Element 3B (General) exam, study these questions in Chapter 11: 3I-8.4, questions 3I-9.3 through 3I-9.5, 3I-12.1 3I-12.2 and 3I-12.5.]

Chapter 10

Element 3A Question Pool

Don't Start Here!

Before you read the questions and answers printed in this chapter, be sure to read the appropriate text in the previous chapters. Use these questions as review exercises when suggested in the text. You should not attempt to memorize all 288 questions and answers in the Element 3A question pool. The material presented in this book has been carefully written and scientifically prepared to guide you step by step through the learning process. By understanding the electronics principles and Amateur Radio concepts as they are presented, your insight into our hobby and your appreciation for the privileges granted by an Amateur Radio license will be greatly enhanced.

This chapter contains the complete question pool for the Technician class license written exam (Element 3A). The FCC specifies the minimum number of questions for an Element 3A exam, and also specifies that a certain percentage of the questions from each subelement must appear on the exam. Most VECs are now giving 25-question Element 3A and 3B exams, and based on that number, there must be five questions from the Rules and Regulations section, subelement 3AA, three questions from the Operating Procedures section, subelement 3AB, and so on. The number of questions to be selected from each section is printed at the beginning of each subelement, and is summarized in Table 10-1.

The FCC now allows Volunteer-Examiner teams to select the questions that will be used on amateur exams. If your test is coordinated by the ARRL/VEC, however, your test will be prepared by the VEC. The multiple choice answers and distractors

Table 10-1
Technician Exam Content

Subelement	3AA	3AB	3AC	3AD	3AE	3AF	3AG	3AH	3AI
Number of Questions	5	3	3	4	2	2	1	2	3

printed here, along with the answer key, have been carefully prepared and evaluated by the ARRL Staff and volunteers in the Field Organization of the League. Most Volunteer-Examiner Coordinators, including the ARRL/VEC, agreed to use the question pool printed here throughout 1987. If your exam is coordinated by the ARRL/VEC or one of the other VECs using this pool, the questions and multiple-choice answers on your test will appear just as they do in this book. Some VECs use the questions printed here with different answers; check with the VEC coordinating your test session.

The Element 3A and Element 3B question pools were created in March 1987 by dividing the original Element 3 question pool. The subelements were renumbered (subelement 3AA and subelement 3BA came from the original subelement 3A, etc.) but the original question numbers were retained. There were two questions numbered 3A-14.3 and two questions numbered 3F-3.5. One question with each of these numbers is now in the Element 3A pool, and one of each is in the Element 3B pool. The FCC also skipped question 3A-14.4 when they made up the original Element 3 question pool, so there is no question with this number in either the Element 3A or the Element 3B pool.

We have listed page references along with the answers in the answer key section of this chapter. These page numbers indicate where you will find the text discussion related to each question. If you have problems with a question, refer back to the page listed with the answer. You may have to study beyond the listed page numbers. There are no page references for the questions in the Rules and Regulations subelements (3AA and 3BA). Reference material for the rules and regulations questions can be found in *The FCC Rule Book*, published by the ARRL.

One final word about the separation of the Element 3 question pool. While you need to study only the questions in Element 3A if you plan to take an examination for a Technician class license, please remember that you are responsible for questions from *both* Element 3A and Element 3B if you plan to upgrade directly from Novice class (or no license) to General class. If you have a Technician license issued prior to March 21, 1987, you automatically receive credit for both Element 3A and Element 3B, and only need to pass the 13 WPM code test to upgrade to General class.

SUBELEMENT 3AA—Rules and Regulations (5 questions)

3A-1.1   What is the *control point* of an amateur station?
    A. The operating position of an Amateur Radio station where the control operator function is performed
    B. The operating position of any Amateur Radio station operating as a repeater user station
    C. The physical location of any Amateur Radio transmitter, even if it is operated by radio link from some other location
    D. The variable frequency oscillator (VFO) of the transmitter

3A-1.2   What is the term for the operating position of an amateur station where the control operator function is performed?
    A. The operating desk
    B. The control point
    C. The station location
    D. The manual control location

3A-2.1   What is an amateur *emergency communication*?
    A. An Amateur Radio communication directly relating to the immediate safety of life of individuals or the immediate protection of property
    B. A communication with the manufacturer of the amateur's equipment in case of equipment failure
    C. The only type of communication allowed in the Amateur Radio Service
    D. A communication that must be left to the Public Safety Radio Services; e.g., police and fire officials

3A-2.2   What is the term for an amateur radiocommunication directly related to the immediate safety of life of an individual?
    A. Immediate safety communication
    B. Emergency communication
    C. Third-party communication
    D. Individual communication

3A-2.3   What is the term for an amateur radiocommunication directly related to the immediate protection of property?
    A. Emergency communication
    B. Immediate communication
    C. Property communication
    D. Priority traffic

3A-2.4   Under what circumstances does the FCC declare that a *general state of communications emergency* exists?
    A. When a declaration of war is received from Congress
    B. When the maximum usable frequency goes above 28 MHz
    C. When communications facilities in Washington, DC, are disrupted
    D. In the event of an emergency disrupting normally available communication facilities in any widespread area(s)

3A-2.5   How does an amateur operator request the FCC to declare that a *general state of communications emergency* exists?
   A. Communication with the FCC engineer-in-charge of the affected area
   B. Communication with the US senator or congressman for the area affected
   C. Communication with the local Emergency Coordinator
   D. Communication with the Chief of the FCC Private Radio Bureau

3A-2.6   What type of instructions are included in an FCC declaration of a *general state of communications emergency*?
   A. Designation of the areas affected and of organizations authorized to use radiocommunications in the affected area
   B. Designation of amateur frequency bands for use only by amateurs participating in emergency communications in the affected area, and complete suspension of Novice operating privileges for the duration of the emergency
   C. Designation of the areas affected and specification of the amateur frequency bands or segments of such bands for use only by amateurs participating in emergency communication within or with such affected area(s)
   D. Suspension of amateur rules regarding station identification and business communication

3A-2.7   What should be done by the control operator of an amateur station which has been designated by the FCC to assist in making known information relating to a *general state of communications emergency*?
   A. The designated station shall act as an official liaison station with local news media and law-enforcement officials
   B. The designated station shall monitor the designated emergency communications frequencies and warn noncomplying stations of the state of emergency
   C. The designated station shall broadcast hourly bulletins from the FCC concerning the disaster situation
   D. The designated station shall coordinate the operation of all phone-patch traffic out of the designated area

3A-2.8   During an FCC-declared *general state of communications emergency*, how must the operation by, and with, amateur stations in the area concerned be conducted?
   A. All transmissions within all designated amateur communications bands other than communications relating directly to relief work, emergency service, or the establishment and maintenance of efficient Amateur Radio networks for the handling of such communications shall be suspended
   B. Operations shall be governed by part 97.93 of the FCC rules pertaining to emergency communications
   C. No amateur operation is permitted in the area during the duration of the declared emergency
   D. Operation by and with amateur stations in the area concerned shall be conducted in the manner the amateur concerned believes most effective to the speedy resolution of the emergency situation

3A-3.1   Notwithstanding the numerical limitations in the FCC Rules, how much transmitting power shall be used by an amateur station?
A. There is no regulation other than the numerical limits
B. The minimum power level required to achieve S9 signal reports
C. The minimum power necessary to carry out the desired communication
D. The maximum power available, as long as it is under the allowable limit

3A-3.6   What is the maximum transmitting power permitted an amateur station in beacon operation?
A. 10 watts PEP output
B. 100 watts PEP output
C. 500 watts PEP output
D. 1500 watts PEP output

3A-3.8   What is the maximum transmitting power permitted an amateur station on 146.52-MHz?
A. 200 watts PEP output
B. 500 watts ERP
C. 1000 watts dc input
D. 1500 watts PEP output

3A-4.2   How must a newly-upgraded Technician control operator with a *Certificate of Successful Completion of Examination* identify the station while it is transmitting on 146.34-MHz pending receipt of a new operator license?
A. The new Technician may not operate on 146.34 until his/her new license arrives
B. The licensee gives his/her call sign, followed by the word "temporary" and the identifier code shown on the certificate of successful completion
C. No special form of identification is needed
D. The licensee gives his/her call sign and states the location of the VE examination where he or she obtained the certificate of successful completion

3A-4.4   Which language(s) must be used when making the station identification by telephony?
A. The language being used for the contact may be used if it is not English, providing the US has a third-party traffic agreement with that country
B. English must be used for identification
C. Any language may be used, if the country which uses that language is a member of the International Telecommunication Union
D. The language being used for the contact must be used for identification purposes

3A-4.5   What aid does the FCC recommend to assist in station identification when using telephony?
A. A speech compressor
B. Q signals
C. An internationally recognized phonetic alphabet
D. Distinctive phonetics, made up by the operator and easy to remember

3A-4.6   What emission mode may always be used for station identification, regardless of the transmitting frequency?
A. A1A
B. F1B
C. A2B
D. A3E

3A-5.1   Under what circumstances, if any, may a third-party participate in radiocommunications from an amateur station?
A. A control operator must be present and continuously monitor and supervise the radio communication to ensure compliance with the rules. In addition, contacts may only be made with amateurs in the US and countries with which the US has a third-party traffic agreement
B. A control operator must be present and continuously monitor and supervise the radio communication to ensure compliance with the rules only if contacts are made with amateurs in countries with which the US has no third-party traffic agreement
C. A control operator must be present and continuously monitor and supervise the radio communication to ensure compliance with the rules. In addition, the control operator must key the transmitter and make the station identification.
D. A control operator must be present and continuously monitor and supervise the radio communication to ensure compliance with the rules. In addition, if contacts are made on frequencies below 30 MHz, the control operator must transmit the call signs of both stations involved in the contact at 10-minute intervals

3A-5.2   Where must the control operator be situated when a third-party is participating in radiocommunications from an amateur station?
A. If a radio remote control is used, the control operator may be physically separated from the control point, when provisions are incorporated to shut off the transmitter by remote control
B. If the control operator supervises the third party until he or she is satisfied of the competence of the third party, the control operator may leave the control point
C. The control operator must stay at the control point for the entire time the third party is participating
D. If the third party holds a valid radiotelegraph license issued by the FCC, no supervision is necessary

3A-5.3   What must the control operator do while a third-party is participating in radiocommunications?
A. If the third party holds a valid commercial radiotelegraph license, no supervision is necessary
B. The control operator must tune up and down 5 kHz from the transmitting frequency on another receiver, to ensure that no interference is taking place
C. If a radio control link is available, the control operator may leave the room
D. The control operator must continuously monitor and supervise the radiocommunication to ensure compliance with the rules

3A-5.4   Under what circumstances, if any, may a third-party assume the duties of the control operator of an amateur station?
A. If the third party holds a valid commercial radiotelegraph license, he or she may act as control operator
B. Under no circumstances may a third party assume the duties of control operator
C. During Field Day, the third party may act as control operator
D. An Amateur Extra class licensee may designate a third party as control operator, if the station is operated above 450 MHz

3A-6.3   What types of material compensation, if any, may be involved in third-party traffic transmitted by an amateur station?
A. Payment of an amount agreed upon by the amateur operator and the parties involved
B. Assistance in maintenance of auxiliary station equipment
C. Donation of amateur equipment to the control operator
D. No compensation may be accepted

3A-6.4   What types of business communications, if any, may be transmitted by an amateur station on behalf of a third-party?
A. Section 97.57 specifically prohibits business communications in the Amateur Service
B. Business communications involving the sale of Amateur Radio equipment
C. Business communications involving an emergency, as defined in Part 97
D. Business communications aiding a broadcast station

3A-6.5   When are third-party messages limited to those of a technical nature relating to tests, and to remarks of a personal character for which, by reason of their unimportance, recourse to the public telecommunications service is not justified?
A. Only when communicating with a person in a country with which the US does not share a third-party traffic agreement
B. When communicating with a non-profit organization such as the ARRL
C. When communicating with the FCC
D. Communications between amateurs in different countries are always limited to those of a technical nature relating to tests and remarks of a personal nature for which, by reason of their unimportance, recourse to the public telecommunications service is not justified

3A-7.1   What kinds of one-way communications by amateur stations are not considered broadcasting?
A. All types of one-way communications by amateurs are considered by the FCC as broadcasting
B. Beacon operation, radio-control operation, emergency communications, information bulletins consisting solely of subject matter relating to Amateur Radio, roundtable discussions and code-practice transmissions
C. Only code-practice transmissions conducted simultaneously on all available amateur bands below 30 MHz and conducted for more than 40 hours per week are not considered broadcasting
D. Only actual emergency communications during a declared communications emergency are exempt

3A-7.2   What is a *one-way radiocommunication*?
          A. A communication in which propagation at the frequency in use
             supports signal travel in only one direction
          B. A communication in which different emissions are used in
             each direction
          C. A communication in which an amateur station transmits to and
             receives from a station in a radio service other than amateur
          D. A transmission to which no on-the-air response is desired or
             expected

3A-7.3   What kinds of one-way information bulletins may be transmitted by
          amateur stations?
          A. NOAA weather bulletins
          B. Commuter traffic reports from local radio stations
          C. Regularly scheduled announcements concerning Amateur
             Radio equipment for sale or trade
          D. Bulletins consisting solely of information relating to Amateur
             Radio

3A-7.4   What types of one-way amateur radiocommunications may be
          transmitted by an amateur station?
          A. Beacon operation, radio control, code practice, retransmission
             of other services
          B. Beacon operation, radio control, transmitting an unmodulated
             carrier, NOAA weather bulletins
          C. Beacon operation, radio control, information bulletins
             consisting solely of information relating to Amateur Radio,
             code practice and emergency communications
          D. Beacon operation, emergency-drill-practice transmissions,
             automatic retransmission of NOAA weather transmissions,
             code practice

3A-8.1   What are the HF privileges authorized to a Technician control
          operator?
          A. 3700 to 3750 kHz, 7100 to 7150 kHz (7050 to 7075 kHz when
             terrestrial station location is in Alaska or Hawaii or outside
             Region 2), 14,100 to 14,150 kHz, 21,100 to 21,150 kHz, and
             28,100 to 28,150 kHz only
          B. 3700 to 3750 kHz, 7100 to 7150 kHz (7050 to 7075 kHz when
             terrestrial station location is in Alaska or Hawaii or outside
             Region 2), 21,100 to 21,200 kHz, and 28,100 to 28,500 kHz
             only
          C. 28,000 to 29,700 kHz only
          D. 3700 to 3750 kHz, 7100 to 7150 kHz (7050 to 7075 kHz when
             terrestrial station location is in Alaska or Hawaii or outside
             Region 2), and 21,100 to 21,200 kHz only

3A-8.2   Which operator licenses authorize privileges on 52.525-MHz?
          A. Extra, Advanced only
          B. Extra, Advanced, General only
          C. Extra, Advanced, General, Technician only
          D. Extra, Advanced, General, Technician, Novice

3A-8.3   Which operator licenses authorize privileges on 146.52-MHz?
          A. Extra, Advanced, General, Technician, Novice
          B. Extra, Advanced, General, Technician only
          C. Extra, Advanced, General only
          D. Extra, Advanced only

3A-8.4   Which operator licenses authorize privileges on 223.50-MHz?
- A. Extra, Advanced, General, Technician, Novice
- B. Extra, Advanced, General, Technician only
- C. Extra, Advanced, General only
- D. Extra, Advanced only

3A-8.5   Which operator licenses authorize privileges on 446.0-MHz?
- A. Extra, Advanced, General, Technician, Novice
- B. Extra, Advanced, General, Technician only
- C. Extra, Advanced, General only
- D. Extra, Advanced only

3A-10.9   On what frequencies within the 6 meter band may emission F3E be transmitted?
- A. 50.0-54.0 MHz only
- B. 50.1-54.0 MHz only
- C. 51.0-54.0 MHz only
- D. 52.0-54.0 MHz only

3A-10.10   On what frequencies within the 2 meter band may emission F3F be transmitted?
- A. 144.1-148.0 MHz only
- B. 146.0-148.0 MHz only
- C. 144.0-148.0 MHz only
- D. 146.0-147.0 MHz only

3A-11.1   What is the nearest to the band edge the transmitting frequency should be set?
- A. 3 kHz for single sideband and 1 kHz for CW
- B. 1 kHz for single sideband and 3 kHz for CW
- C. 1.5 kHz for single sideband and 0.05 kHz for CW
- D. As near as the operator desires, providing that no sideband, harmonic, or spurious emission (in excess of that legally permitted) falls outside the band

3A-11.2   When selecting the transmitting frequency, what allowance should be made for sideband emissions resulting from keying or modulation?
- A. The sidebands must be adjacent to the authorized Amateur Radio frequency band in use
- B. The sidebands must be harmonically-related frequencies that fall outside of the Amateur Radio frequency band in use
- C. The sidebands must be confined within the authorized Amateur Radio frequency band occupied by the carrier
- D. The sidebands must fall outside of the Amateur Radio frequency band in use so as to prevent interference to other Amateur Radio stations

3A-12.1   What is the maximum mean output power an amateur station is permitted in order to operate under the special rules for radio control of remote model craft and vehicles?
- A. One watt
- B. One milliwatt
- C. Two watts
- D. Three watts

3A-12.2   What information must be indicated on the writing affixed to the transmitter in order to operate under the special rules for radio control of remote model craft and vehicles?
A. Station call sign
B. Station call sign and operating times
C. Station call sign and licensee's name and address
D. Station call sign, class of license, and operating times

3A-12.3   What are the station identification requirements for an amateur station operated under the special rules for radio control of remote model craft and vehicles?
A. Once every ten minutes, and at the beginning and end of each transmission
B. Once every ten minutes
C. At the beginning and end of each transmission
D. Station identification is not required

3A-12.4   Where must the writing indicating the station call sign and the licensee's name and address be affixed in order to operate under the special rules for radio control of remote model craft and vehicles?
A. It must be in the operator's possession
B. It must be affixed to the transmitter
C. It must be affixed to the craft or vehicle
D. It must be filed with the nearest FCC Field Office

3A-13.3   What is the maximum sending speed permitted for an emission F1B transmission between 28- and 50-MHz?
A. 56 kilobauds
B. 19.6 kilobauds
C. 1200 bauds
D. 300 bauds

3A-13.4   What is the maximum sending speed permitted for an emission F1B transmission between 50- and 220-MHz?
A. 56 kilobauds
B. 19.6 kilobauds
C. 1200 bauds
D. 300 bauds

3A-13.5   What is the maximum sending speed permitted for an emission F1B transmission above 220-MHz?
A. 300 bauds
B. 1200 bauds
C. 19.6 kilobauds
D. 56 kilobauds

3A-13.6   What is the maximum frequency shift permitted for emission F1B when transmitted below 50-MHz?
A. 100 Hz
B. 500 Hz
C. 1000 Hz
D. 5000 Hz

3A-13.7   What is the maximum frequency shift permitted for emission F1B when transmitted above 50-MHz?
A. 100 Hz or the sending speed, in bauds, whichever is greater
B. 500 Hz or the sending speed, in bauds, whichever is greater
C. 1000 Hz or the sending speed, in bauds, whichever is greater
D. 5000 Hz or the sending speed, in bauds, whichever is greater

3A-13.8   What is the maximum bandwidth permitted an amateur station
           transmission between 50- and 220-MHz using a non-standard digital
           code?
           A. 20 kHz
           B. 50 kHz
           C. 80 kHz
           D. 100 kHz

3A-13.9   What is the maximum bandwidth permitted an amateur station
           transmission between 220- and 902-MHz using a non-standard
           digital code?
           A. 20 kHz
           B. 50 kHz
           C. 80 kHz
           D. 100 kHz

3A-13.10   What is the maximum bandwidth permitted an amateur station
           transmission above 902-MHz using a non-standard digital code?
           A. 20 kHz
           B. 100 kHz
           C. 200 kHz, as defined by Section 97.66 (g)
           D. Any bandwidth, providing that the emission is in accordance
              with section 97.63 (b) and 97.73 (c)

3A-14.1   What is meant by the term *broadcasting*?
           A. The dissemination of radio communications intended to be
              received by the public directly or by intermediary relay stations
           B. Retransmission by automatic means of programs or signals
              emanating from any class of station other than amateur
           C. The transmission of any one-way radio communication,
              regardless of purpose or content   •
           D. Any one-way or two-way radio communication involving more
              than two stations

3A-14.2   What classes of station may be automatically retransmitted by an
           amateur station?
           A. FCC licensed commercial stations
           B. Federally or state-authorized Civil Defense stations
           C. Amateur Radio stations
           D. National Weather Service bulletin stations

3A-14.3   Under what circumstances, if any, may a broadcast station
           retransmit the signals from an amateur station?
           A. Under no circumstances
           B. When the amateur station is not used for any activity directly
              related to program production or newsgathering for broadcast
              purposes
           C. If the station rebroadcasting the signal feels that such action
              would benefit the public
           D. When no other forms of communication exist

**3A-14.4 THIS QUESTION WAS SKIPPED WHEN FCC MADE UP THE
           QUESTION POOL. THERE IS NO QUESTION WITH THIS
           NUMBER.

3A-14.5   Under what circumstances, if any, may an amateur station
           retransmit a NOAA weather station broadcast?
           A. If the NOAA broadcast is taped and retransmitted later
           B. If a general state of communications emergency is declared by
              the FCC
           C. If permission is granted by NOAA for amateur retransmission
              of the broadcast
           D. Under no circumstances

3A-14.7   Under what circumstances, if any, may an amateur station be used
           for an activity related to program production or news-gathering for
           broadcast purposes?
           A. The programs or news produced with the assistance of an
              amateur station must be taped for broadcast at a later time
           B. An amateur station may be used for newsgathering and
              program production only by National Public Radio
           C. Under no circumstances
           D. Programs or news produced with the assistance of an amateur
              station must mention the call sign of that station

3A-15.2   Under what circumstances, if any, may singing be transmitted by an
           amateur station?
           A. When the singing produces no dissonances or spurious
              emissions
           B. When it is used to jam an illegal transmission
           C. Only above 1215 MHz
           D. Transmitting music is not permitted in the Amateur Service

3A-17.1   Under what circumstances, if any, may an amateur station transmit
           radiocommunications containing obscene words?
           A Obscene words are permitted when they do not cause
              interference to any other radio communication or signal
           B. Obscene words are prohibited in Amateur Radio transmissions
           C. Obscene words are permitted when they are not retransmitted
              through repeater or auxiliary stations
           D. Obscene words are permitted, but there is an unwritten rule
              among amateurs that they should not be used on the air

3A-17.2   Under what circumstances, if any, may an amateur station transmit
           radiocommunications containing indecent words?
           A. Indecent words are permitted when they do not cause
              interference to any other radio communication or signal
           B. Indecent words are permitted when they are not retransmitted
              through repeater or auxiliary stations
           C. Indecent words are permitted, but there is an unwritten rule
              among amateurs that they should not be used on the air
           D. Indecent words are prohibited in Amateur Radio transmissions

3A-17.3   Under what circumstances, if any, may an amateur station transmit
           radiocommunications containing profane words?
           A. Profane words are permitted when they are not retransmitted
              through repeater or auxiliary stations
           B. Profane words are permitted, but there is an unwritten rule
              among amateurs that they should not be used on the air
           C. Profane words are prohibited in Amateur Radio transmissions
           D. Profane words are permitted when they do not cause
              interference to any other radio communication or signal

SUBELEMENT 3AB—Operating Procedures (3 questions)

3B-1.1   What is the meaning of: "Your report is five seven..."?
A. Your signal is perfectly readable and moderately strong
B. Your signal is perfectly readable, but weak
C. Your signal is readable with considerable difficulty
D. Your signal is perfectly readable with near pure tone

3B-1.2   What is the meaning of: "Your report is three three..."?
A. Your signal is readable with considerable difficulty and weak in strength
B. The station is located at latitude 33 degrees
C. The contact is serial number thirty-three
D. Your signal is unreadable, very weak in strength

3B-1.3   What is the meaning of: "Your report is five nine plus 20 dB..."?
A. Your signal strength has increased by a factor of 100
B. Repeat your transmission on a frequency 20 kHz higher
C. The bandwidth of your signal is 20 decibels above linearity
D. A relative signal-strength meter reading is 20 decibels greater than strength 9

3B-1.6   How should the microphone gain control be adjusted on an emission F3E transmitter?
A. For proper deviation on modulation peaks
B. For maximum, non-clipped amplitude on modulation peaks
C. For moderate movement of the ALC meter on modulation peaks
D. For a dip in plate current

3B-1.7   How is the call sign WE5TZD stated phonetically?
A. Whiskey-Echo-Foxtrot-Tango-Zulu-Delta
B. Washington-England-Five-Tokyo-Zanzibar-Denmark
C. Whiskey-Echo-Five-Tango-Zulu-Delta
D. Whiskey-Easy-Five-Tear-Zebra-Dog

3B-1.8   How is the call sign KC4HRM stated phonetically?
A. Kilo-Charlie-Four-Hotel-Romeo-Mike
B. Kilowatt-Charlie-Four-Hotel-Roger-Mexico
C. Kentucky-Canada-Four-Honolulu-Radio-Mexico
D. King-Charlie-Foxtrot-Hotel-Roger-Mary

3B-1.9   How is the call sign AF6PSQ stated phonetically?
A. America-Florida-Six-Portugal-Spain-Quebec
B. Adam-Frank-Six-Peter-Sugar-Queen
C. Alfa-Fox-Sierra-Papa-Santiago-Queen
D. Alfa-Foxtrot-Six-Papa-Sierra-Quebec

3B-1.10   How is the call sign NB8LXG stated phonetically?
A. November-Bravo-Eight-Lima-Xray-Golf
B. Nancy-Baker-Eight-Love-Xray-George
C. Norway-Boston-Eight-London-Xray-Germany
D. November-Bravo-Eight-London-Xray-Germany

3B-1.11   How is the call sign KJ1UOI stated phonetically?
A. King-John-One-Uncle-Oboe-Ida
B. Kilowatt-George-India-Uncle-Oscar-India
C. Kilo-Juliette-One-Uniform-Oscar-India
D. Kentucky-Juliette-One-United-Ontario-Indiana

3B-1.12   How is the call sign WV2BPZ stated phonetically?
           A. Whiskey-Victor-Two-Bravo-Papa-Zulu
           B. Willie-Victor-Two-Baker-Papa-Zebra
           C. Whiskey-Victor-Tango-Bravo-Papa-Zulu
           D. Willie-Virginia-Two-Boston-Peter-Zanzibar

3B-1.13   How is the call sign NY3CTJ stated phonetically?
           A. Norway-Yokohama-Three-California-Tokyo-Japan
           B. Nancy-Yankee-Three-Cat-Texas-Jackrabbit
           C. Norway-Yesterday-Three-Charlie-Texas-Juliette
           D. November-Yankee-Three-Charlie-Tango-Juliette

3B-1.14   How is the call sign KG7DRV stated phonetically?
           A. Kilo-Golf-Seven-Denver-Radio-Venezuela
           B. Kilo-Golf-Seven-Delta-Romeo-Victor
           C. King-John-Seven-Dog-Radio-Victor
           D. Kilowatt-George-Seven-Delta-Romeo-Video

3B-1.15   How is the call sign WX9HKS stated phonetically?
           A. Whiskey-Xray-Nine-Hotel-Kilo-Sierra
           B. Willie-Xray-November-Hotel-King-Sierra
           C. Washington-Xray-Nine-Honolulu-Kentucky-Santiago
           D. Whiskey-Xray-Nine-Henry-King-Sugar

3B-1.16   How is the call sign AE0LQY stated phonetically?
           A. Able-Easy-Zero-Lima-Quebec-Yankee
           B. Arizona-Equador-Zero-London-Queen-Yesterday
           C. Alfa-Echo-Zero-Lima-Quebec-Yankee
           D. Able-Easy-Zero-Love-Queen-Yoke

3B-2.5   What is meant by the term *AMTOR*?
           A. AMTOR is a system using two separate antennas with a
              common receiver to reduce transmission errors
           B. AMTOR is a system in which the transmitter feeds two
              antennas, at right angles to each other, to reduce transmission
              errors
           C. AMTOR is a system using independent sideband to reduce
              transmission errors
           D. AMTOR is a system using error-detection and correction to
              reduce transmission errors

3B-2.7   What is the most common frequency shift for emission F2B
           transmissions in the amateur VHF bands?
           A. 85 Hz
           B. 170 Hz
           C. 300 Hz
           D. 425 Hz

3B-2.8   What is an *RTTY mailbox*?
           A. A QSL Bureau for teletype DX cards
           B. An open net for RTTY operators
           C. An address to which RTTY operators may write for technical
              assistance
           D. A system by which messages may be stored electronically for
              later retrieval

3B-2.9   What is the purpose of transmitting a string of *RYRYRY* characters in RTTY?
A. It is the RTTY equivalent of CQ
B. Since it represents alternate upper and lower case signals, it is used to assist the receiving operator check the shift mechanism
C. Since it contains alternating mark and space frequencies, it is a check on proper operation of the transmitting and receiving equipment
D. It is sent at the beginning of an important message to activate stations equipped with SELCAL and Autostart

3B-3.1   How should a QSO be initiated through a station in repeater operation?
A. Say "breaker, breaker 79"
B. Call the desired station and then identify your own station
C. Call "CQ" three times and identify three times
D. Wait for a "CQ" to be called and then answer it

3B-3.2   Why should users of a station in repeater operation pause briefly between transmissions?
A. To check the SWR of the repeater
B. To reach for pencil and paper for third party traffic
C. To listen for any hams wanting to break in
D. To dial up the repeater's autopatch

3B-3.3   Why should users of a station in repeater operation keep their transmissions short and thoughtful?
A. A long transmission may prevent someone with an emergency from using the repeater
B. To see if the receiving station operator is still awake
C. To give any non-hams that are listening a chance to respond
D. To keep long-distance charges down

3B-3.4   Why should simplex be used where possible instead of using a station in repeater operation?
A. Farther distances can be reached
B. To avoid long distance toll charges
C. To avoid tying up the repeater unnecessarily
D. To permit the testing of the effectiveness of your antenna

3B-3.5   What is the proper procedure to break into an on-going QSO through a station in repeater operation?
A. Wait for the end of a transmission and start calling
B. Shout, "break, break!" to show that you're eager to join the conversation
C. Turn on your 100-watt amplifier and override whoever is talking
D. Send your call sign during a break between transmissions

3B-3.6   What is the purpose of repeater operation?
A. To cut your power bill by using someone's higher power system
B. To enable mobile and low-power stations to extend their usable range
C. To reduce your telephone bill
D. To call the ham radio distributor 50 miles away

3B-3.7   What is a *repeater frequency coordinator*?
           A. Someone who coordinates the assembly of a repeater station
           B. Someone who provides advice on what kind of system to buy
           C. The club's repeater trustee
           D. A person or group that recommends frequency pairs for
              repeater usage

3B-3.9   What is the usual input/output frequency separation for stations in
           repeater operation in the 2 meter band?
           A. 1 MHz
           B. 1.6 MHz
           C. 170 Hz
           D. 0.6 MHz

3B-3.10   What is the usual input/output frequency separation for stations in
           repeater operation in the 70 centimeter band?
           A. 1.6 MHz
           B. 5 MHz
           C. 600 kHz
           D. 5 kHz

3B-3.11   What is the usual input/output frequency separation for a 6 meter
           station in repeater operation?
           A. 1 MHz
           B. 600 kHz
           C. 1.6 MHz
           D. 20 kHz

3B-3.13   What is the usual input/output frequency separation for a 1.25 meter
           station in repeater operation?
           A. 1000 kHz
           B. 600 kHz
           C. 1600 kHz
           D. 1.6 GHz

3B-6.4   Why should local amateur radiocommunications be conducted on
           VHF and UHF frequencies?
           A. To minimize interference on HF bands capable of long-
              distance sky-wave communication
           B. Because greater output power is permitted on VHF and UHF
           C. Because HF transmissions are not propagated locally
           D. Because absorption is greater at VHF and UHF frequencies

3B-6.5   How can on-the-air transmissions be minimized during a lengthy
           transmitter testing or loading up procedure?
           A. Use a dummy antenna
           B. Choose an unoccupied frequency
           C. Use a non-resonant antenna
           D. Use a resonant antenna that requires no loading up procedure

3B-6.6   When a frequency conflict arises between a simplex operation and a
           repeater operation, why does good amateur practice call for the
           simplex operation to move to another frequency?
           A. The repeater's output power can be turned up to ruin the front
              end of the station in simplex operation
           B. There are more repeaters than simplex operators
           C. Changing the repeater's frequency is not practical
           D. Changing a repeater frequency requires the authorization of
              the Federal Communications Commission

3B-6.7   What should be done before installing an amateur station within one mile of an FCC monitoring station?
A. The amateur should apply to the FCC for a Special Temporary Authority for operation within the shadow of the monitoring facility's antenna system
B. The amateur should make sure a line-of-sight path does not exist between the amateur station and the monitoring facility
C. The amateur should consult with the Commission to protect the monitoring facility from harmful interference
D. The amateur should make sure the effective radiated power of the amateur station will be less than 200 watts PEP in the direction of the monitoring facility

3B-6.8   What is the proper Q signal to use to determine whether a frequency is in use before making a transmission?
A. QRL?
B. QRU?
C. QRV?
D. QRZ?

3B-6.9   What is meant by "making the repeater time out"?
A. The repeater's battery supply has run out
B. The repeater's transmission time limit has expired during a single transmission
C. The warranty on the repeater duplexer has expired
D. The repeater is in need of repairs

3B-6.10   During commuting rush hours, which types of operation should relinquish the use of the repeater?
A. Mobile operators
B. Low-power stations
C. Highway traffic information nets
D. Third-party traffic nets

3B-9.1   What is the proper distress calling procedure when using telephony?
A. Transmit MAYDAY
B. Transmit QRRR
C. Transmit $\overline{QRZ}$
D. Transmit $\overline{SOS}$

3B-9.2   What is the proper distress calling procedure when using telegraphy?
A. Transmit MAYDAY
B. Transmit QRRR
C. Transmit QRZ
D. Transmit $\overline{SOS}$

SUBELEMENT 3AC—Radio-Wave Propagation (3 questions)

3C-1.1   What is the *ionosphere*?
A. That part of the upper atmosphere where enough ions and free electrons exist to affect radio-wave propagation
B. The boundary between two air masses of different temperature and humidity, along which radio waves can travel
C. The ball that goes on the top of a mobile whip antenna
D. That part of the atmosphere where weather takes place

3C-1.2   Which ionospheric layer limits daytime radiocommunications in the 80 meter band to short distances?
A. D layer
B. F1 layer
C. E layer
D. F2 layer

3C-1.3   What is the region of the outer atmosphere which makes long-distance radiocommunications possible as a result of bending of radio waves?
A. Troposphere
B. Stratosphere
C. Magnetosphere
D. Ionosphere

3C-1.4   Which layer of the ionosphere is mainly responsible for long-distance sky-wave radiocommunications?
A. D layer
B. E layer
C. F1 layer
D. F2 layer

3C-1.5   What are the two distinct sub-layers of the F layer of the ionosphere during the daytime?
A. Troposphere and stratosphere
B. F1 and F2
C. Electrostatic and electromagnetic
D. D and E

3C-1.8   What is the lowest region of the ionosphere that is useful for long-distance radio wave propagation?
A. The D layer
B. The E layer
C. The F1 layer
D. The F2 layer

3C-1.11 What type of solar radiation is most responsible for ionization in the outer atmosphere?
A. Thermal
B. Ionized particle
C. Ultraviolet
D. Microwave

3C-1.12 What is the lowest ionospheric layer?
A. The A layer
B. The D layer
C. The E layer
D. The F layer

3C-1.14   What is the region of the outer atmosphere which makes long-distance radiocommunications possible as a result of bending of the radio waves?
A. The ionosphere
B. The troposphere
C. The magnetosphere
D. The stratosphere

3C-2.1   Which layer of the ionosphere is most responsible for absorption of radio signals during daylight hours?
A. The E layer
B. The F1 layer
C. The F2 layer
D. The D layer

3C-2.2   When is ionospheric absorption most pronounced?
A. When radio waves enter the D layer at low angles
B. When tropospheric ducting occurs
C. When radio waves travel to the F layer
D. When a temperature inversion occurs

3C-2.5   During daylight hours, what effect does the D layer of the ionosphere have on 80 meter radio waves?
A. The D layer absorbs the signals
B. The D layer bends the radio waves out into space
C. The D layer refracts the radio waves back to earth
D. The D layer has little or no effect on 80 meter radio wave propagation

3C-2.6   What causes *ionospheric absorption* of radio waves?
A. A lack of D layer ionization
B. D layer ionization
C. The presence of ionized clouds in the E layer
D. Splitting of the F layer

3C-3.1   What is the highest radio frequency that will be refracted back to earth called?
A. Lowest usable frequency
B. Optimum working frequency
C. Ultra high frequency
D. Critical frequency

3C-3.2   What causes the *maximum usable frequency* to vary?
A. Variations in the temperature of the air at ionospheric levels
B. Upper-atmospheric wind patterns
C. Presence of ducting
D. The amount of ultraviolet and other types of radiation received from the sun

3C-3.5   What does the term *maximum usable frequency* refer to?
A. The maximum frequency that allows a radio signal to reach its destination in a single hop
B. The minimum frequency that allows a radio signal to reach its destination in a single hop
C. The maximum frequency that allows a radio signal to be absorbed in the lowest ionospheric layer
D. The minimum frequency that allows a radio signal to be absorbed in the lowest ionospheric layer

3C-4.1   What is usually the condition of the ionosphere just before sunrise?
A. Atmospheric attenuation is at a maximum
B. Ionization is at a maximum
C. The E layer is above the F layer
D. Ionization is at a minimum

3C-4.2   At what time of day does maximum ionization of the ionosphere occur?
A. Dusk
B. Midnight
C. Dawn
D. Midday

3C-4.3   Which two daytime ionospheric layers combine into one layer at night?
A. E and F1
B. D and E
C. E1 and E2
D. F1 and F2

3C-4.4   Minimum ionization of the ionosphere occurs daily at what time?
A. Shortly before dawn
B. Just after noon
C. Just after dusk
D. Shortly before midnight

3C-6.1   When two stations are within each other's skip zone on the frequency being used, what mode of propagation would it be desirable to use?
A. Ground wave propagation
B. Sky wave propagation
C. Scatter-mode propagation
D. Ionospheric ducting propagation

3C-6.3   When is E layer ionization at a maximum?
A. Dawn
B. Midday
C. Dusk
D. Midnight

3C-8.1   What is the transmission path of a wave that travels directly from the transmitting antenna to the receiving antenna called?
A. Line of sight
B. The sky wave
C. The linear wave
D. The plane wave

3C-8.2   How are VHF signals within the range of the visible horizon propagated?
A. By sky wave
B. By direct wave
C. By plane wave
D. By geometric wave

3C-9.1   Ducting occurs in which region of the atmosphere?
A. F2
B. Ionosphere
C. Stratosphere
D. Troposphere

3C-9.2   What effect does tropospheric bending have on 2 meter radio waves?
- A. It increases the distance over which they can be transmitted
- B. It decreases the distance over which they can be transmitted
- C. It tends to garble 2-meter phone transmissions
- D. It reverses the sideband of 2-meter phone transmissions

3C-9.3   What atmospheric phenomenon causes tropospheric ducting of radio waves?
- A. A very low pressure area
- B. An aurora to the north
- C. Lightning between the transmitting and receiving station
- D. A temperature inversion

3C-9.4   Tropospheric ducting occurs as a result of what phenomenon?
- A. A temperature inversion
- B. Sun spots
- C. An aurora to the north
- D. Lightning between the transmitting and receiving station

3C-9.5   What atmospheric phenomenon causes VHF radio waves to be propagated several hundred miles through stable air masses over oceans?
- A. Presence of a maritime polar air mass
- B. A widespread temperature inversion
- C. An overcast of cirriform clouds
- D. Atmospheric pressure of roughly 29 inches of mercury or higher

3C-9.6   In what frequency range does tropospheric ducting occur most often?
- A. LF
- B. MF
- C. HF
- D. VHF

SUBELEMENT 3AD—Amateur Radio Practice (4 questions)

3D-1.1   Where should the green wire in an ac line cord be attached in a power supply?
A. To the fuse
B. To the "hot" side of the power switch
C. To the chassis
D. To the meter

3D-1.2   Where should the black (or red) wire in a three-wire line cord be attached in a power supply?
A. To the filter capacitor
B. To the dc ground
C. To the chassis
D. To the fuse

3D-1.3   Where should the white wire in a three-wire line cord be attached in a power supply?
A. To the fuse
B. To one side of the transformer's primary winding
C. To the black wire
D. To the rectifier junction

3D-1.4   Why is the retaining screw in one terminal of a light socket made of brass while the other one is silver colored?
A. To prevent galvanic action
B. To indicate correct polarity
C. To better conduct current
D. To reduce skin effect

3D-2.1   How much electrical current flowing through the human body is usually fatal?
A. As little as 100 milliamperes may be fatal
B. Approximately 10 amperes is required to be fatal
C. More than 20 amperes is needed to kill a human being
D. No amount of current will harm you. Voltages of over 2000 volts are always fatal, however

3D-2.2   What is the minimum voltage considered to be dangerous to humans?
A. 30 volts
B. 100 volts
C. 1000 volts
D. 2000 volts

3D-2.3   Where should the main power-line switch for a high voltage power supply be situated?
A. Inside the cabinet, to interrupt power when the cabinet is opened
B. On the rear panel of the high-voltage supply
C. Where it can be seen and reached easily
D. This supply should not be switch-operated

3D-2.5   How much electrical current flowing through the human body is usually painful?
A. As little as 50 milliamperes may be painful
B. Approximately 10 amperes is required to be painful
C. More than 20 amperes is needed to be painful to a human being
D. No amount of current will be painful. Voltages of over 2000 volts are always painful, however

3D-5.2   Where in the antenna transmission line should a peak-reading
           wattmeter be attached to determine the transmitter output power?
           A. At the transmitter output
           B. At the antenna feed point
           C. One-half wavelength from the antenna feed point
           D. One-quarter wavelength from the transmitter output

3D-5.3   If a directional rf wattmeter indicates 90 watts forward power and 10
           watts reflected power, what is the actual transmitter output power?
           A. 10 watts
           B. 80 watts
           C. 90 watts
           D. 100 watts

3D-5.4   If a directional rf wattmeter indicates 96 watts forward power and 4
           watts reflected power, what is the actual transmitter output power?
           A. 80 watts
           B. 88 watts
           C. 92 watts
           D. 100 watts

3D-7.1   What is a *multimeter*?
           A. An instrument capable of reading SWR and power
           B. An instrument capable of reading resistance, capacitance, and
              inductance
           C. An instrument capable of reading resistance and reactance
           D. An instrument capable of reading voltage, current, and
              resistance

3D-7.2   How can the range of a voltmeter be extended?
           A. By adding resistance in series with the circuit under test
           B. By adding resistance in parallel with the circuit under test
           C. By adding resistance in series with the meter
           D. By adding resistance in parallel with the meter

3D-7.3   How is a voltmeter typically connected to a circuit under test?
           A. In series with the circuit
           B. In parallel with the circuit
           C. In quadrature with the circuit
           D. In phase with the circuit

3D-7.4   How can the range of an ammeter be extended?
           A. By adding resistance in series with the circuit under test
           B. By adding resistance in parallel with the circuit under test
           C. By adding resistance in series with the meter
           D. By adding resistance in parallel with the meter

3D-8.1   What is a *marker generator*?
           A. A high-stability oscillator that generates a series of reference
              signals at known frequency intervals
           B. A low-stability oscillator that "sweeps" through a band of
              frequencies
           C. An oscillator often used in aircraft to determine the craft's
              location relative to the inner and outer markers at airports
           D. A high-stability oscillator whose output frequency and
              amplitude can be varied over a wide range

3D-8.2   What piece of test equipment provides a variable-frequency signal which can be used to check the frequency response of a circuit?
A. Frequency counter
B. Distortion analyzer
C. Deviation meter
D. Signal generator

3D-8.3   What type of circuit is used to inject a frequency calibration signal into a communication receiver?
A. A product detector
B. A receiver incremental tuning circuit
C. A balanced modulator
D. A crystal calibrator

3D-8.4   How is a *marker generator* used?
A. To calibrate the tuning dial on a receiver
B. To calibrate the volume control on a receiver
C. To test the amplitude linearity of an SSB transmitter
D. To test the frequency deviation of an FM transmitter

3D-8.5   When adjusting a transmitter filter circuit, what device is connected to the transmitter output?
A. Multimeter
B. Litz wires
C. Receiver
D. Dummy antenna

3D-11.1   What is a *reflectometer*?
A. An instrument used to measure signals reflected from the ionosphere
B. An instrument used to measure standing wave ratio
C. An instrument used to measure transmission-line impedance
D. An instrument used to measure radiation resistance

3D-11.2   For best accuracy when adjusting the impedance match between an antenna and feed line, where should the match-indicating device be inserted?
A. At the antenna feed point
B. At the transmitter
C. At the midpoint of the feed line
D. Anywhere along the feed line

3D-11.3   What is the device that can indicate an impedance mismatch in an antenna system?
A. A field-strength meter
B. A set of lecher wires
C. A wavemeter
D. A reflectometer

3D-11.4   What is a *reflectometer*?
A. An instrument used to measure signals reflected from the ionosphere
B. An instrument used to measure standing wave ratio
C. An instrument used to measure transmission-line impedance
D. An instrument used to measure radiation resistance

3D-11.5   Where should a reflectometer be inserted into a long antenna
           transmission line in order to obtain the most valid standing wave
           ratio indication?
           A. At any quarter-wavelength interval along the transmission line
           B. At the receiver end
           C. At the antenna end
           D. At any even half-wavelength interval along the transmission
              line

3D-12.1   What result might be expected when using a speech processor with
           an emission J3E transmitter?
           A. A lower plate-current reading
           B. A less natural-sounding voice
           C. A cooler operating power supply
           D. Greater PEP output

3D-14.1   What is a *transmatch*?
           A. A device for varying the resonant frequency of an antenna
           B. A device for varying the impedance presented to the
              transmitter
           C. A device for varying the tuning rate of the transmitter
           D. A device for varying the electrical length of an antenna

3D-14.2   What is a *balanced line*?
           A. Feed line with one conductor connected to ground
           B. Feed line with both conductors connected to ground to
              balance out harmonics
           C. Feed line with the outer conductor connected to ground at
              even intervals
           D. Feed line with neither conductor connected to ground

3D-14.3   What is an *unbalanced line*?
           A. Feed line with neither conductor connected to ground
           B. Feed line with both conductors connected to ground to
              suppress harmonics
           C. Feed line with one conductor connected to ground
           D. Feed line with the outer conductor connected to ground at
              uneven intervals

3D-14.4   What is a *balun*?
           A. A device for using an unbalanced line to supply power to a
              balanced load, or vice versa
           B. A device to match impedances between two coaxial lines
           C. A device used to connect a microphone to a balanced
              modulator
           D. A counterbalance used with an azimuth/elevation rotator
              system

3D-14.5   What is the purpose of an antenna matching circuit?
           A. To measure the impedance of the antenna
           B. To compare the radiation patterns of two antennas
           C. To measure the SWR of an antenna
           D. To match impedances within the antenna system

3D-14.8   How is a *transmatch* used?
A. It is connected between a transmitter and an antenna system, and tuned for minimum SWR at the transmitter
B. It is connected between a transmitter and an antenna system and tuned for minimum SWR at the antenna
C. It is connected between a transmitter and an antenna system, and tuned for minimum impedance
D. It is connected between a transmitter and a dummy load, and tuned for maximum output power

3D-16.1   What is a *dummy antenna*?
A. An isotropic radiator
B. A nonradiating load for a transmitter
C. An antenna used as a reference for gain measurements
D. The image of an antenna, located below ground

3D-16.2   Of what materials may a dummy antenna be made?
A. A wire-wound resistor
B. A noninductive resistor
C. A diode and resistor combination
D. A coil and capacitor combination

3D-16.3   What station accessory is used in place of an antenna during transmitter tests so that no signal is radiated?
A. A Transmatch
B. A dummy antenna
C. A low-pass filter
D. A decoupling resistor

3D-16.4   What is the purpose of a *dummy load*?
A. To allow off-the-air transmitter testing
B. To reduce output power for QRP operation
C. To give comparative signal reports
D. To allow Transmatch tuning without causing interference

3D-16.5   How many watts should a dummy load for use with a 100 watt emission J3E transmitter with 50 ohm output be able to dissipate?
A. A minimum of 100 watts continuous
B. A minimum of 141 watts continuous
C. A minimum of 175 watts continuous
D. A minimum of 200 watts continuous

3D-17.1   What is an *S-meter*?
A. A meter used to measure sideband suppression
B. A meter used to measure spurious emissions from a transmitter
C. A meter used to measure relative signal strength in a receiver
D. A meter used to measure solar flux

3D-18.1   For the most accurate readings of transmitter output power, where should the rf wattmeter be inserted?
A. The wattmeter should be inserted and the output measured one-quarter wavelength from the antenna feed point
B. The wattmeter should be inserted and the output measured one-half wavelength from the antenna feed point
C. The wattmeter should be inserted and the output power measured at the transmitter antenna jack
D. The wattmeter should be inserted and the output power measured at the Transmatch output

3D-18.2   At what line impedance are rf wattmeters usually designed to
           operate?
           A. 25 ohms
           B. 50 ohms
           C. 100 ohms
           D. 300 ohms

3D-18.3   What is a *directional wattmeter*?
           A. An instrument that measures forward or reflected power
           B. An instrument that measures the directional pattern of an
              antenna
           C. An instrument that measures the energy consumed by the
              transmitter
           D. An instrument that measures thermal heating in a load resistor

SUBELEMENT 3AE—Electrical Principles (2 questions)

3E-2.1   What is meant by the term *resistance*?
  A. The opposition to the flow of current in an electric circuit containing inductance
  B. The opposition to the flow of current in an electric circuit containing capacitance
  C. The opposition to the flow of current in an electric circuit containing reactance
  D. The opposition to the flow of current in an electric circuit that does not contain reactance

3E-2.2   What is the primary function of a resistor?
  A. To store an electric charge
  B. To store a magnetic field
  C. To match a high-impedance source to a low-impedance load
  D. To limit the current in an electric circuit

3E-2.3   What is a *variable resistor*?
  A. A resistor with a slide or contact that makes the resistance adjustable
  B. A device that can transform a variable voltage into a constant voltage
  C. A resistor that changes value when an ac voltage is applied to it
  D. A resistor that changes value when it is heated

3E-2.4   Why do resistors generate heat?
  A. They convert electrical energy to heat energy
  B. They exhibit reactance
  C. Because of skin effect
  D. To produce thermionic emission

3E-4.1   What is an *inductor*?
  A. An electronic component that stores energy in an electric field
  B. An electronic component that converts a high voltage to a lower voltage
  C. An electronic component that opposes dc while allowing ac to pass
  D. An electronic component that stores energy in a magnetic field

3E-4.2   What factors determine the amount of inductance in a coil?
  A. The type of material used in the core, the diameter of the core and whether the coil is mounted horizontally or vertically
  B. The diameter of the core, the number of turns of wire used to wind the coil and the type of metal used in the wire
  C. The type of material used in the core, the number of turns used to wind the core and the frequency of the current through the coil
  D. The type of material used in the core, the diameter of the core, the length of the coil and the number of turns of wire used to wind the coil

3E-4.3   What are the electrical properties of an inductor?
    A. An inductor stores a charge electrostatically and opposes a change in voltage
    B. An inductor stores a charge electrochemically and opposes a change in current
    C. An inductor stores a charge electromagnetically and opposes a change in current
    D. An inductor stores a charge electromechanically and opposes a change in voltage

3E-4.4   What is an inductor *core*?
    A. The central portion of a coil; may be made from air, iron, brass or other material
    B. A tight coil of wire used in a transformer
    C. An insulating material placed between the plates of an inductor
    D. The point at which an inductor is tapped to produce resonance

3E-4.5   What are the component parts of a coil?
    A. The wire in the winding and the core material
    B. Two conductive plates and an insulating material
    C. Two or more layers of silicon material
    D. A donut-shaped iron core and a layer of insulating tape

3E-5.1   What is a *capacitor*?
    A. An electronic component that stores energy in a magnetic field
    B. An electronic component that stores energy in an electric field
    C. An electronic component that converts a high voltage to a lower voltage
    D. An electronic component that converts power into heat

3E-5.2   What factors determine the amount of capacitance in a capacitor?
    A. The dielectric constant of the material between the plates, the area of one side of one plate, the separation between the plates and the number of plates
    B. The dielectric constant of the material between the plates, the number of plates and the diameter of the leads connected to the plates
    C. The number of plates, the spacing between the plates and whether the dielectric material is N type or P type
    D. The dielectric constant of the material between the plates, the surface area of one side of one plate, the number of plates and the type of material used for the protective coating

3E-5.3   What are the electrical properties of a capacitor?
    A. A capacitor stores a charge electrochemically and opposes a change in current
    B. A capacitor stores a charge electromagnetically and opposes a change in current
    C. A capacitor stores a charge electromechanically and opposes a change in voltage
    D. A capacitor stores a charge electrostatically and opposes a change in voltage

3E-5.4   What is a capacitor *dielectric*?
    A. The insulating material used for the plates
    B. The conducting material used between the plates
    C. The ferrite material that the plates are mounted on
    D. The insulating material between the plates

3E-5.5   What are the component parts of a capacitor?
           A. Two or more conductive plates with an insulating material
              between them
           B. The wire used in the winding and the core material
           C. Two or more layers of silicon material
           D. Two insulating plates with a conductive material between them

3E-7.1   What is an *ohm*?
           A. The basic unit of resistance
           B. The basic unit of capacitance
           C. The basic unit of inductance
           D. The basic unit of admittance

3E-7.3   What is the unit measurement of resistance?
           A. Volt
           B. Ampere
           C. Joule
           D. Ohm

3E-8.1   What is a *microfarad*?
           A. A basic unit of capacitance equal to 10^{-6} farads
           B. A basic unit of capacitance equal to 10^{-12} farads
           C. A basic unit of capacitance equal to 10^{-2} farads
           D. A basic unit of capacitance equal to 10^{6} farads

3E-8.2   What is a *picofarad*?
           A. A basic unit of capacitance equal to 10^{-6} farads
           B. A basic unit of capacitance equal to 10^{-12} farads
           C. A basic unit of capacitance equal to 10^{-2} farads
           D. A basic unit of capacitance equal to 10^{6} farads

3E-8.3   What is a *farad*?
           A. The basic unit of resistance
           B. The basic unit of capacitance
           C. The basic unit of inductance
           D. The basic unit of admittance

3E-8.4   What is the basic unit of capacitance?
           A. Farad
           B. Ohm
           C. Volt
           D. Ampere

3E-9.1   What is a *microhenry*?
           A. A basic unit of inductance equal to 10^{-12} henrys
           B. A basic unit of inductance equal to 10^{-3} henrys
           C. A basic unit of inductance equal to 10^{6} henrys
           D. A basic unit of inductance equal to 10^{-6} henrys

3E-9.2   What is a *millihenry*?
           A. A basic unit of inductance equal to 10^{-6} henrys
           B. A basic unit of inductance equal to 10^{-12} henrys
           C. A basic unit of inductance equal to 10^{-3} henrys
           D. A basic unit of inductance equal to 10^{6} henrys

3E-9.3   What is a *henry*?
           A. The basic unit of resistance
           B. The basic unit of capacitance
           C. The basic unit of inductance
           D. The basic unit of admittance

3E-9.4   What is the basic unit of inductance?
- A. Coulomb
- B. Farad
- C. Ohm
- D. Henry

3E-11.1   How is the current in a dc circuit calculated when the voltage and resistance are known?
- A. $I = E / R$
- B. $P = I \times E$
- C. $I = R \times E$
- D. $I = E \times R$

3E-11.2   What is the input resistance of a load when a 12-volt battery supplies 0.25-amperes to it?
- A. 0.02 ohms
- B. 3 ohms
- C. 48 ohms
- D. 480 ohms

3E-11.3   The product of the current and what force gives the electrical power in a circuit?
- A. Magnetomotive force
- B. Centripetal force
- C. Electrochemical force
- D. Electromotive force

3E-11.4   What is *Ohm's Law*?
- A. A mathematical relationship between resistance, current and applied voltage in a circuit
- B. A mathematical relationship between current, resistance and power in a circuit
- C. A mathematical relationship between current, voltage and power in a circuit
- D. A mathematical relationship between resistance, voltage and power in a circuit

3E-11.5   What is the input resistance of a load when a 12-volt battery supplies 0.15-amperes to it?
- A. 8 ohms
- B. 80 ohms
- C. 100 ohms
- D. 800 ohms

3E-12.2   In a series circuit composed of a voltage source and several resistors, what determines the voltage drop across any particular resistor?
- A. It is equal to the source voltage
- B. It is equal to the source voltage divided by the number of series resistors in the circuit
- C. The larger the resistor's value, the greater the voltage drop across that resistor
- D. The smaller the resistor's value, the greater the voltage drop across that resistor

3E-13.4   How is power calculated when the current and voltage in a circuit are known?
- A. $E = I \times R$
- B. $P = I \times E$
- C. $P = I^2 / R$
- D. $P = E / I$

3E-14.8   When 120-volts is measured across a 4700 ohm resistor, approximately how much current is flowing through it?
A. 39 amperes
B. 3.9 amperes
C. 0.26 ampere
D. 0.026 ampere

3E-14.9   When 120-volts is measured across a 47000 ohm resistor, approximately how much current is flowing through it?
A. 392 A
B. 39.2 A
C. 26 mA
D. 2.6 mA

3E-14.10   When 12-volts is measured across a 4700 ohm resistor, approximately how much current is flowing through it?
A. 2.6 mA
B. 26 mA
C. 39.2 A
D. 392 A

3E-14.11   When 12-volts is measured across a 47000 ohm resistor, approximately how much current is flowing through it?
A. 255 μA
B. 255 mA
C. 3917 mA
D. 3917 A

SUBELEMENT 3AF—Circuit Components (2 questions)

3F-1.1   How can a carbon resistor's electrical tolerance rating be found?
A. By using a wavemeter
B. By using the resistor's color code
C. By using Thevenin's theorem for resistors
D. By using the Baudot code

3F-1.2   Why would a large size resistor be substituted for a smaller one of the same resistance?
A. To obtain better response
B. To obtain a higher current gain
C. To increase power dissipation capability
D. To produce a greater parallel impedance

3F-1.3   What do the first three color bands on a resistor indicate?
A. The value of the resistor in ohms
B. The resistance tolerance in percent
C. The power rating in watts
D. The value of the resistor in henrys

3F-1.4   What does the fourth color band on a resistor indicate?
A. The value of the resistor in ohms
B. The resistance tolerance in percent
C. The power rating in watts
D. The resistor composition

3F-1.6   When the color bands on a group of resistors indicate that they all have the same resistance, what further information about each resistor is needed in order to select those that have nearly equal value?
A. The working voltage rating of each resistor
B. The composition of each resistor
C. The tolerance of each resistor
D. The current rating of each resistor

3F-2.1   As the plate area of a capacitor is increased, what happens to its capacitance?
A. Decreases
B. Increases
C. Stays the same
D. Becomes voltage dependent

3F-2.2   As the plate spacing of a capacitor is increased, what happens to its capacitance?
A. Increases
B. Stays the same
C. Becomes voltage dependent
D. Decreases

3F-2.3   What is an *electrolytic capacitor*?
A. A capacitor whose plates are formed on a thin ceramic layer
B. A capacitor whose plates are separated by a thin strip of mica insulation
C. A capacitor whose dielectric is formed on one set of plates through electrochemical action
D. A capacitor whose value varies with applied voltage

3F-2.4   What is a *paper capacitor*?
A. A capacitor whose plates are formed on a thin ceramic layer
B. A capacitor whose plates are separated by a thin strip of mica insulation
C. A capacitor whose plates are separated by a layer of paper
D. A capacitor whose dielectric is formed on one set of plates through electrochemical action

3F-2.5   What factors must be considered when selecting a capacitor for a circuit?
A. Type of capacitor, capacitance and voltage rating
B. Type of capacitor, capacitance and the kilowatt-hour rating
C. The amount of capacitance, the temperature coefficient and the KVA rating
D. The type of capacitor, the microscopy coefficient and the temperature coefficient

3F-2.8   How are the characteristics of a capacitor usually specified?
A. In volts and amperes
B. In microfarads and volts
C. In ohms and watts
D. In millihenrys and amperes

3F-3.1   What can be done to raise the inductance of a 5-microhenry air-core coil to a 5-millihenry coil with the same physical dimensions?
A. The coil can be wound on a non-conducting tube
B. The coil can be wound on an iron core
C. Both ends of the coil can be brought around to form the shape of a donut, or toroid
D. The coil can be made of a heavier-gauge wire

3F-3.2   Describe an *inductor*.
A. A semiconductor in a conducting shield
B. Two parallel conducting plates
C. A straight wire conductor mounted inside a Faraday shield
D. A coil of conducting wire

3F-3.3   As an iron core is inserted in a coil, what happens to the inductance?
A. It increases
B. It decreases
C. It stays the same
D. It becomes voltage-dependent

3F-3.4   As a brass core is inserted in a coil, what happens to the inductance?
A. It increases
B. It decreases
C. It stays the same
D. It becomes voltage-dependent

3F-3.5   For radio frequency power applications, which type of inductor has the least amount of loss?
A. Magnetic wire
B. Iron core
C. Air core
D. Slug tuned

3F-3.6   Where does an inductor store energy?
          A. In a capacitive field
          B. In a magnetic field
          C. In an electrical field
          D. In a resistive field

3F-5.3   What is a *heat sink*?
          A. A device used to heat an electrical component uniformly
          B. A device used to remove heat from an electronic component
          C. A tub in which circuit boards are soldered
          D. A fan used for transmitter cooling

SUBELEMENT 3AG—Practical Circuits (1 question)

3G-2.1   What is a high-pass filter usually connected to?
    A. The transmitter and the Transmatch
    B. The Transmatch and the transmission line
    C. The television receiving antenna and a television receiver's antenna input
    D. The transmission line and the transmitting antenna

3G-2.2   Where is the proper place to install a high-pass filter?
    A. At the antenna terminals of a television receiver
    B. Between a transmitter and a Transmatch
    C. Between a Transmatch and the transmission line
    D. On a transmitting antenna

3G-2.3   Where is a band-pass filter usually installed?
    A. Between the spark plugs and coil in a mobile setup
    B. On a transmitting antenna
    C. In a communications receiver
    D. Between a Transmatch and the transmitting antenna

3G-2.4   Which frequencies are attenuated by a low-pass filter?
    A. Those above its cut-off frequency
    B. Those within its cut-off frequency
    C. Those within 50 kHz on either side of its cut-off frequency
    D. Those below its cut-off frequency

3G-2.5   What circuit passes electrical energy above a certain frequency and attenuates electrical energy below that frequency?
    A. An input filter
    B. A low-pass filter
    C. A high-pass filter
    D. A band-pass filter

3G-2.6   What circuit passes electrical energy below a certain frequency and blocks electrical energy above that frequency?
    A. An input filter
    B. A low-pass filter
    C. A high-pass filter
    D. A band-pass filter

3G-2.7   What circuit attenuates electrical energy above a certain frequency and below a lower frequency?
    A. An input filter
    B. A low-pass filter
    C. A high-pass filter
    D. A band-pass filter

3G-2.9   What general range of rf energy does a band-pass filter reject?
    A. All frequencies above a specified frequency
    B. All frequencies below a specified frequency
    C. All frequencies above the upper limit of the band in question
    D. All frequencies above a specified frequency and below a lower specified frequency

3G-3.1   What circuit is likely to be found in all types of receivers?
    A. A detector
    B. An RF amplifier
    C. An audio filter
    D. A beat frequency oscillator

3G-3.2   In a filter-type emission J3E transmitter, what stage combines rf and af energy to produce a double-sideband suppressed carrier signal?
A. The product detector
B. The automatic-load-control circuit
C. The balanced modulator
D. The local oscillator

3G-3.3   In a superheterodyne receiver for emission A3E reception, what stage combines the received rf with energy from the local oscillator to produce a signal at the receiver intermediate frequency?
A. The mixer
B. The detector
C. The RF amplifier
D. The AF amplifier

SUBELEMENT 3AH—Signals and Emissions (2 questions)

3H-1.1   What is emission type NØN?
- A. Unmodulated carrier
- B. Telegraphy by on-off keying
- C. Telegraphy by keyed tone
- D. Telegraphy by frequency-shift keying

3H-1.2   What is emission type A3E?
- A. Frequency-modulated telephony
- B. Facsimile
- C. Double-sideband, amplitude-modulated telephony
- D. Amplitude-modulated telegraphy

3H-1.3   What is emission type J3E?
- A. Single-sideband suppressed-carrier amplitude-modulated telephony
- B. Single-sideband suppressed-carrier amplitude-modulated telegraphy
- C. Independent sideband suppressed-carrier amplitude-modulated telephony
- D. Single-sideband suppressed-carrier frequency-modulated telephony

3H-1.4   What is emission type F1B?
- A. Amplitude-shift-keyed telegraphy
- B. Frequency-shift-keyed telegraphy
- C. Frequency-modulated telephony
- D. Phase-modulated telephony

3H-1.5   What is emission type F2B?
- A. Frequency-modulated telephony
- B. Frequency-modulated telegraphy using audio tones
- C. Frequency-modulated facsimile using audio tones
- D. Phase-modulated television

3H-1.6   What is emission type F3E?
- A. AM telephony
- B. AM telegraphy
- C. FM telegraphy
- D. FM telephony

3H-1.7   What is the emission symbol for telegraphy by frequency shift keying without the use of a modulating tone?
- A. F1B
- B. F2B
- C. A1A
- D. J3E

3H-1.8   What is the emission symbol for telegraphy by the on-off keying of a frequency modulated tone?
- A. F1B
- B. F2A
- C. A1A
- D. J3E

3H-1.9   What is the emission symbol for telephony by amplitude
          modulation?
          A. A1A
          B. A3E
          C. J2B
          D. F3E

3H-1.10   What is the emission symbol for telephony by frequency
          modulation?
          A. F2B
          B. F3E
          C. A3E
          D. F1B

3H-2.2   What is the meaning of the term *modulation*?
          A. The process of varying some characteristic of a carrier wave
             for the purpose of conveying information
          B. The process of recovering audio information from a received
             signal
          C. The process of increasing the average power of a single-
             sideband transmission
          D. The process of suppressing the carrier in a single-sideband
             transmitter

3H-6.1   What characteristic makes emission F3E especially well-suited for
          local VHF/UHF radiocommunications?
          A. Good audio fidelity and intelligibility under weak-signal
             conditions
          B. Good audio fidelity and high signal-to-noise ratio above a
             certain signal amplitude threshold
          C. Better rejection of multipath distortion than the AM modes
          D. Better carrier frequency stability than the AM modes

3H-6.2   What emission is produced by a transmitter using a reactance
          modulator?
          A. A1A
          B. NØN
          C. J3E
          D. G3E

3H-7.1   What other emission does phase modulation most resemble?
          A. Amplitude modulation
          B. Pulse modulation
          C. Frequency modulation
          D. Single-sideband modulation

3H-9.2   What emission does not have sidebands resulting from modulation?
          A. A3E
          B. NØN
          C. F3E
          D. F2B

3H-12.1   To what is the deviation of an emission F3E transmission
          proportional?
          A. Only the frequency of the audio modulating signal
          B. The frequency and the amplitude of the audio modulating
             signal
          C. The duty cycle of the audio modulating signal
          D. Only the amplitude of the audio modulating signal

3H-14.1   What is the result of overdeviation in an emission F3E transmitter?
           A. Increased transmitter power consumption
           B. Out-of-channel emissions (splatter)
           C. Increased transmitter range
           D. Inadequate carrier suppression

3H-14.2   What is *splatter*?
           A. Interference to adjacent signals caused by excessive
              transmitter keying speeds
           B. Interference to adjacent signals caused by improper transmitter
              neutralization
           C. Interference to adjacent signals caused by overmodulation of a
              transmitter
           D. Interference to adjacent signals caused by parasitic oscillations
              at the antenna

3H-16.1   What emissions are used in teleprinting?
           A. F1A, F2B and F1B
           B. A2B, F1B and F2B
           C. A1B, A2B and F2B
           D. A2B, F1A and F2B

3H-16.2   What two states of teleprinter codes are most commonly used in
           amateur radiocommunications?
           A. Dot and dash
           B. Highband and lowband
           C. Start and stop
           D. Mark and space

3H-16.3   What emission type results when an af shift keyer is connected to
           the microphone jack of an emission F3E transmitter?
           A. A2B
           B. F1B
           C. F2B
           D. A1F

SUBELEMENT 3AI—Antennas and Feed Lines (3 questions)

3I-1.1   What antenna type best strengthens signals from a particular
          direction while attenuating those from other directions?
          A. A monopole antenna
          B. An isotropic antenna
          C. A vertical antenna
          D. A beam antenna

3I-1.2   What is a *Yagi* antenna?
          A. Half-wavelength elements stacked vertically and excited in
             phase
          B. Quarter-wavelength elements arranged horizontally and excited
             out of phase
          C. Half-wavelength linear driven element(s) with parasitically
             excited parallel linear elements
          D. Quarter-wavelength, triangular loop elements

3I-1.4   What is the general configuration of the radiating elements of a
          horizontally-polarized Yagi?
          A. Two or more straight, parallel elements arranged in the same
             horizontal plane
          B. Vertically stacked square or circular loops arranged in parallel
             horizontal planes
          C. Two or more wire loops arranged in parallel vertical planes
          D. A vertical radiator arranged in the center of an effective RF
             ground plane

3I-1.5   What type of parasitic beam antenna uses two or more straight
          metal-tubing elements arranged physically parallel to each other?
          A. A quad antenna
          B. A delta loop antenna
          C. A Zepp antenna
          D. A Yagi antenna

3I-1.6   How many directly-driven elements does a Yagi antenna have?
          A. None; they are all parasitic
          B. One
          C. Two
          D. All elements are directly driven

3I-1.8   What is a *parasitic beam antenna*?
          A. An antenna where the director and reflector elements receive
             their RF excitation by induction or radiation from the driven
             element
          B. An antenna where wave traps are used to assure magnetic
             coupling among the elements
          C. An antenna where all elements are driven by direct connection
             to the feed line
          D. An antenna where the driven element receives its RF
             excitation by induction or radiation from the directors

3I-2.2   What kind of antenna array is composed of a square full-wave
          closed loop driven element with parallel parasitic element(s)?
          A. Dual rhombic
          B. Cubical quad
          C. Stacked Yagi
          D. Delta loop

3I-2.3   Approximately how long is one side of the driven element of a cubical quad antenna?
- A. 2 electrical wavelengths
- B. 1 electrical wavelength
- C. ½ electrical wavelength
- D. ¼ electrical wavelength

3I-2.4   Approximately how long is the wire in the driven element of a cubical quad antenna?
- A. ¼ electrical wavelength
- B. ½ electrical wavelength
- C. 1 electrical wavelength
- D. 2 electrical wavelengths

3I-2.5   What is a *delta loop antenna*?
- A. A variation of the cubical quad antenna, with triangular elements
- B. A large copper ring, used in direction finding
- C. An antenna system composed of three vertical antennas, arranged in a triangular shape
- D. An antenna made from several coils of wire on an insulating form

3I-2.6   What is a *cubical quad antenna*?
- A. Four parallel metal tubes, each approximately ½ electrical wavelength long
- B. Two or more parallel four-sided wire loops, each approximately one electrical wavelength long
- C. A vertical conductor ¼ electrical wavelength high, fed at the bottom
- D. A center-fed wire ½ electrical wavelength long

3I-4.1   What is the polarization of electromagnetic waves radiated from a half-wavelength antenna perpendicular to the earth's surface?
- A. Circularly polarized waves
- B. Horizontally polarized waves
- C. Parabolically polarized waves
- D. Vertically polarized waves

3I-4.2   What is the electromagnetic wave polarization of most man-made electrical noise radiation in the HF-VHF spectrum?
- A. Left-hand circular
- B. Vertical
- C. Right-hand circular
- D. Horizontal

3I-4.3   To what does the term *vertical* as applied to wave polarization refer?
- A. This means that the electric lines of force in the radio wave are parallel to the earth's surface
- B. This means that the magnetic lines of force in the radio wave are perpendicular to the earth's surface
- C. This means that the electric lines of force in the radio wave are perpendicular to the earth's surface
- D. This means that the radio wave will leave the antenna and radiate vertically into the ionosphere

3I-4.4   To what does the term *horizontal* as applied to wave polarization refer?
    A. This means that the magnetic lines of force in the radio wave are parallel to the earth's surface
    B. This means that the electric lines of force in the radio wave are parallel to the earth's surface
    C. This means that the electric lines of force in the radio wave are perpendicular to the earth's surface
    D. This means that the radio wave will leave the antenna and radiate horizontally to the destination

3I-4.5   What electromagnetic wave polarization does a cubical quad antenna have when the feed point is in the center of a horizontal side?
    A. Vertical
    B. Horizontal
    C. Circular
    D. Helical

3I-4.6   What electromagnetic wave polarization does a cubical quad antenna have when the feed point is in the center of a vertical side?
    A. Vertical
    B. Horizontal
    C. Circular
    D. Helical

3I-4.7   What electromagnetic wave polarization does a cubical quad antenna have when all sides are at 45 degrees to the earth's surface and the feed point is at the bottom corner?
    A. Vertical
    B. Horizontal
    C. Circular
    D. Helical

3I-4.8   What electromagnetic wave polarization does a cubical quad antenna have when all sides are at 45 degrees to the earth's surface and the feed point is at a side corner?
    A. Vertical
    B. Horizontal
    C. Circular
    D. Helical

3I-6.7   What is a *directional antenna*?
    A. An antenna whose parasitic elements are all constructed to be directors
    B. An antenna that radiates in direct line-of-sight propagation, but not skywave or skip propagation
    C. An antenna permanently mounted so as to radiate in only one direction
    D. An antenna that radiates more strongly in some directions than others

3I-8.1   What is meant by the term *standing wave ratio*?
    A. The ratio of forward and reflected inductances on a feed line
    B. The ratio of forward and reflected resistances on a feed line
    C. The ratio of forward and reflected impedances on a feed line
    D. The ratio of forward and reflected voltages on a feed line

3I-8.2   What is meant by the term *forward power*?
          A. The power traveling from the transmitter to the antenna
          B. The power radiated from the front of a directional antenna
          C. The power produced during the positive half of the RF cycle
          D. The power used to drive a linear amplifier

3I-8.3   What is meant by the term *reflected power*?
          A. The power radiated from the back of a directional antenna
          B. The power returned to the transmitter from the antenna
          C. The power produced during the negative half of the RF cycle
          D. Power reflected to the transmitter site by buildings and trees

3I-9.1   What is *standing wave ratio* a measure of?
          A. The ratio of maximum to minimum voltage on a feed line
          B. The ratio of maximum to minimum reactance on a feed line
          C. The ratio of maximum to minimum resistance on a feed line
          D. The ratio of maximum to minimum sidebands on a feed line

3I-9.2   What happens to the power loss in an unbalanced feed line as the
          standing wave ratio increases?
          A. It is unpredictable
          B. It becomes nonexistent
          C. It decreases
          D. It increases

3I-10.1   What is a *balanced line*?
          A. Feed line with one conductor connected to ground
          B. Feed line with both conductors connected to ground to
             balance out harmonics
          C. Feed line with the outer conductor connected to ground at
             even intervals
          D. Feed line with neither conductor connected to ground

3I-10.2   What is a *balanced antenna*?
          A. A symmetrical antenna with one side of the feed point
             connected to ground
          B. An antenna (or a driven element in a array) that is symmetrical
             about the feed point
          C. A symmetrical antenna with both sides of the feed point
             connected to ground, to balance out harmonics
          D. An antenna designed to be mounted in the center

3I-10.3   What is an *unbalanced line*?
          A. Feed line with neither conductor connected to ground
          B. Feed line with both conductors connected to ground to
             suppress harmonics
          C. Feed line with one conductor connected to ground
          D. Feed line with the outer conductor connected to ground at
             uneven intervals

3I-10.4   What is an *unbalanced antenna*?
          A. An antenna (or a driven element in an array) that is not
             symmetrical about the feed point
          B. A symmetrical antenna, having neither half connected to
             ground
          C. An antenna (or a driven element in an array) that is symmetrical
             about the feed point
          D. A symmetrical antenna with both halves coupled to ground at
             uneven intervals

3I-11.3   What type of feed line is best suited to operating at a high standing wave ratio?
   A. Coaxial cable
   B. Twisted pair
   C. Flat ribbon "twin lead"
   D. Parallel open-wire line

3I-11.5   What is the general relationship between frequencies passing through a feed line and the losses in the feed line?
   A. Loss is independent of frequency
   B. Loss increases with increasing frequency
   C. Loss decreases with increasing frequency
   D. There is no predictable relationship

3I-11.6   What happens to rf energy not delivered to the antenna by a lossy coaxial cable?
   A. It is radiated by the feed line
   B. It is returned to the transmitter's chassis ground
   C. Some of it is dissipated as heat in the conductors and dielectric
   D. It is canceled because of the voltage ratio of forward power to reflected power in the feed line

3I-11.9   As the operating frequency decreases, what happens to conductor losses in a feed line?
   A. The losses decrease
   B. The losses increase
   C. The losses remain the same
   D. The losses become infinite

3I-11.11   As the operating frequency increases, what happens to conductor losses in a feed line?
   A. The losses decrease
   B. The losses increase
   C. The losses remain the same
   D. The losses decrease to zero

3I-12.3   What device can be installed on a balanced antenna so that it can be fed through a coaxial cable?
   A. A triaxial transformer
   B. A wavetrap
   C. A loading coil
   D. A balun

3I-12.4   What is a *balun*?
   A. A device that can be used to convert an antenna designed to be fed at the center so that it may be fed at one end
   B. A device that may be installed on a balanced antenna so that it may be fed with unbalanced feed line
   C. A device that can be installed on an antenna to produce horizontally polarized or vertically polarized waves
   D. A device used to allow an antenna to operate on more than one band

Element 3A Answer Key

SUBELEMENT 3AA

3A-1.1	A
3A-1.2	B
3A-2.1	A
3A-2.2	B
3A-2.3	A
3A-2.4	D
3A-2.5	A
3A-2.6	C
3A-2.7	B
3A-2.8	A
3A-3.1	C
3A-3.6	B
3A-3.8	D
3A-4.2	B
3A-4.4	B
3A-4.5	C
3A-4.6	A
3A-5.1	A
3A-5.2	C
3A-5.3	D
3A-5.4	B
3A-6.3	D
3A-6.4	C
3A-6.5	D
3A-7.1	B
3A-7.2	D
3A-7.3	D
3A-7.4	C
3A-8.1	B
3A-8.2	C
3A-8.3	B
3A-8.4	A
3A-8.5	B
3A-10.9	B
3A-10.10	A
3A-11.1	D
3A-11.2	C
3A-12.1	A
3A-12.2	C
3A-12.3	D
3A-12.4	B
3A-13.3	C
3A-13.4	B

3A-13.5	D
3A-13.6	C
3A-13.7	C
3A-13.8	A
3A-13.9	D
3A-13.10	D
3A-14.1	A
3A-14.2	C
3A-14.3	B
3A-14.4	**
3A-14.5	D
3A-14.7	C
3A-15.2	D
3A-17.1	B
3A-17.2	D
3A-17.3	C

SUBELEMENT 3AB

3B-1.1	A	p.2-3
3B-1.2	A	p.2-3
3B-1.3	D	p.2-3
3B-1.6	A	p.2-2
3B-1.7	C	p.2-2
3B-1.8	A	p.2-2
3B-1.9	D	p.2-2
3B-1.10	A	p.2-2
3B-1.11	C	p.2-2
3B-1.12	A	p.2-2
3B-1.13	D	p.2-2
3B-1.14	B	p.2-2
3B-1.15	A	p.2-2
3B-1.16	C	p.2-2
3B-2.5	D	p.2-8
3B-2.7	B	p.2-7
3B-2.8	D	p.2-8
3B-2.9	C	p.2-7
3B-3.1	B	p.2-4
3B-3.2	C	p.2-4
3B-3.3	A	p.2-4
3B-3.4	C	p.2-4
3B-3.5	D	p.2-4
3B-3.6	B	p.2-4
3B-3.7	D	p.2-5
3B-3.9	D	p.2-6

**This number was skipped when the FCC made up the question pool. There is no question number 3A-14.4.

3B-3.10	B	p.2-6
3B-3.11	A	p.2-6
3B-3.13	C	p.2-6
3B-6.4	A	p.2-12
3B-6.5	A	p.2-12
3B-6.6	C	p.2-12
3B-6.7	C	p.2-12
3B-6.8	A	p.2-9
3B-6.9	B	p.2-12
3B-6.10	D	p.2-12
3B-9.1	A	p.2-16
3B-9.2	D	p.2-16

SUBELEMENT 3AC

3C-1.1	A	p.3-1
3C-1.2	A	p.3-2
3C-1.3	D	p.3-2
3C-1.4	D	p.3-2
3C-1.5	B	p.3-2
3C-1.8	B	p.3-3
3C-1.11	C	p.3-2
3C-1.12	B	p.3-2
3C-1.14	A	p.3-3
3C-2.1	D	p.3-3
3C-2.2	A	p.3-3
3C-2.5	A	p.3-3
3C-2.6	B	p.3-3
3C-3.1	D	p.3-5
3C-3.2	D	p.3-5
3C-3.5	A	p.3-5
3C-4.1	D	p.3-3
3C-4.2	D	p.3-3
3C-4.3	D	p.3-2
3C-4.4	A	p.3-3
3C-6.1	C	p.3-9
3C-6.3	B	p.3-3
3C-8.1	A	p.3-11
3C-8.2	B	p.3-11
3C-9.1	D	p.3-11
3C-9.2	A	p.3-11
3C-9.3	D	p.3-11
3C-9.4	A	p.3-11
3C-9.5	B	p.3-11
3C-9.6	D	p.3-11

SUBELEMENT 3AD

3D-1.1	C	p.4-1
3D-1.2	D	p.4-1
3D-1.3	B	p.4-1
3D-1.4	B	p.4-1
3D-2.1	A	p.4-5
3D-2.2	A	p.4-5
3D-2.3	C	p.4-5

3D-2.5	A	p.4-5
3D-5.2	A	p.8-7
3D-5.3	B	p.9-18
3D-5.4	C	p.9-18
3D-7.1	D	p.4-8
3D-7.2	C	p.4-8
3D-7.3	B	p.4-8
3D-7.4	D	p.4-8
3D-8.1	A	p.4-10
3D-8.2	D	p.4-10
3D-8.3	D	p.4-10
3D-8.4	A	p.4-10
3D-8.5	D	p.4-10
3D-11.1	B	p.4-19
3D-11.2	A	p.4-18
3D-11.3	D	p.4-19
3D-11.4	B	p.4-19
3D-11.5	C	p.4-19
3D-12.1	B	p.4-16
3D-14.1	B	p.4-18
3D-14.2	D	p.4-18
3D-14.3	C	p.4-18
3D-14.4	A	p.4-18
3D-14.5	D	p.4-18
3D-14.8	A	p.4-18
3D-16.1	B	p.4-15
3D-16.2	B	p.4-15
3D-16.3	B	p.4-15
3D-16.4	A	p.4-15
3D-16.5	A	p.4-15
3D-17.1	C	p.4-15
3D-18.1	C	p.4-19
3D-18.2	B	p.4-19
3D-18.3	A	p.4-19

SUBELEMENT 3AE

3E-2.1	D	p.5-4
3E-2.2	D	p.5-4
3E-2.3	A	p.5-5
3E-2.4	A	p.5-5
3E-4.1	D	p.5-15
3E-4.2	D	p.5-17
3E-4.3	C	p.5-15
3E-4.4	A	p.5-15
3E-4.5	A	p.5-15
3E-5.1	B	p.5-13
3E-5.2	A	p.5-14
3E-5.3	D	p.5-13
3E-5.4	D	p.5-14
3E-5.5	A	p.5-14
3E-7.1	A	p.5-14
3E-7.3	D	p.5-14
3E-8.1	A	p.5-14

3E-8.2	B	p.5-14
3E-8.3	B	p.5-14
3E-8.4	A	p.5-14
3E-9.1	D	p.5-17
3E-9.2	C	p.5-17
3E-9.3	C	p.5-17
3E-9.4	D	p. 5-17
3E-11.1	A	p.5-6
3E-11.2	C	p.5-6
3E-11.3	D	p.5-9
3E-11.4	A	p.5-6
3E-11.5	B	p.5-6
3E-12.2	C	p.5-5
3E-13.4	B	p.5-9
3E-14.8	D	p.5-6
3E-14.9	D	p.5-6
3E-14.10	A	p.5-6
3E-14.11	A	p.5-6

SUBELEMENT 3AF

3F-1.1	B	p.6-1
3F-1.2	C	p.6-1
3F-1.3	A	p.6-3
3F-1.4	B	p.6-3
3F-1.6	C	p.6-3
3F-2.1	B	p.6-4
3F-2.2	D	p.6-4
3F-2.3	C	p.6-7
3F-2.4	C	p.6-7
3F-2.5	A	p.6-4
3F-2.8	B	p.6-4
3F-3.1	B	p.6-9
3F-3.2	D	p.6-9
3F-3.3	A	p.6-9
3F-3.4	B	p.6-9
3F-3.5	C	p.6-9
3F-3.6	B	p.6-9
3F-5.3	B	p.6-15

SUBELEMENT 3AG

3G-2.1	C	p.7-8
3G-2.2	A	p.7-8
3G-2.3	C	p.7-8
3G-2.4	A	p.7-7
3G-2.5	C	p.7-8
3G-2.6	B	p.7-7
3G-2.7	D	p.7-8
3G-2.9	D	p.7-8
3G-3.1	A	p.7-9
3G-3.2	C	p.7-14
3G-3.3	A	p.7-10

SUBELEMENT 3AH

3H-1.1	A	p.8-1
3H-1.2	C	p.8-1
3H-1.3	A	p.8-1
3H-1.4	B	p.8-1
3H-1.5	B	p.8-1
3H-1.6	D	p.8-1
3H-1.7	A	p.8-1
3H-1.8	B	p.8-1
3H-1.9	B	p.8-1
3H-1.10	B	p.8-1
3H-2.2	A	p.8-2
3H-6.1	B	p.8-8
3H-6.2	D	p.8-10
3H-7.1	C	p.8-10
3H-9.2	B	p.8-4
3H-12.1	D	p.8-11
3H-14.1	B	p.8-13
3H-14.2	C	p.8-13
3H-16.1	B	p.8-14
3H-16.2	D	p.8-14
3H-16.3	C	p.8-14

SUBELEMENT 3AI

3I-1.1	D	p.9-3
3I-1.2	C	p.9-5
3I-1.4	A	p.9-6
3I-1.5	D	p.9-6
3I-1.6	B	p.9-6
3I-1.8	A	p.9-5
3I-2.2	B	p.9-9
3I-2.3	D	p.9-9
3I-2.4	C	p.9-9
3I-2.5	A	p.9-9
3I-2.6	B	p.9-9
3I-4.1	D	p.9-1
3I-4.2	B	p.9-1
3I-4.3	C	p.9-1
3I-4.4	B	p.9-1
3I-4.5	B	p.9-9
3I-4.6	A	p.9-9
3I-4.7	B	p.9-9
3I-4.8	A	p.9-9
3I-6.7	D	p.9-5
3I-8.1	D	p.9-16
3I-8.2	A	p.9-16
3I-8.3	B	p.9-16
3I-9.1	A	p.9-16
3I-9.2	D	p.9-16
3I-10.1	D	p.9-18
3I-10.2	B	p.9-18
3I-10.3	C	p.9-18
3I-10.4	A	p.9-18

Chapter 11

Element 3B Question Pool

Don't Start Here!

B
efore you read the questions and answers printed in this chapter, be sure to read the appropriate text in the previous chapters. Use these questions as review exercises when suggested in the text. You should not attempt to memorize all 286 questions and answers in the Element 3B question pool. The material presented in this book has been carefully written and scientifically prepared to guide you step by step through the learning process. By understanding the electronics principles and Amateur Radio concepts as they are presented, your insight into our hobby and your appreciation for the privileges granted by an Amateur Radio license will be greatly enhanced.

This chapter contains the complete question pool for the General class license written exam (Element 3B). The FCC specifies the minimum number of questions for an Element 3B exam, and also specifies that a certain percentage of the questions from each subelement must appear on the exam. Most VECs are now giving 25-question Element 3A and 3B exams, and based on that number, there must be four questions from the Rules and Regulations section, subelement 3BA, three questions from the Operating Procedures section, subelement 3BB, and so on. The number of questions to be selected from each section is printed at the beginning of each subelement, and is summarized in Table 11-1.

The FCC now allows Volunteer-Examiner teams to select the questions that will be used on amateur exams. If your test is coordinated by the ARRL/VEC, however,

Table 11-1
General Exam Content

Subelement	3BA	3BB	3BC	3BD	3BE	3BF	3BG	3BH	3BI
Number of Questions	4	3	3	5	2	1	1	2	4

your test will be prepared by the VEC. The multiple choice answers and distractors printed here, along with the answer key, have been carefully prepared and evaluated by the ARRL Staff and volunteers in the Field Organization of the League. Most Volunteer-Examiner Coordinators, including the ARRL/VEC, agreed to use the question pool printed here throughout 1987. If your exam is coordinated by the ARRL/VEC or one of the other VECs using this pool, the questions and multiple-choice answers on your test will appear just as they do in this book. Some VECs use the questions printed here with different answers; check with the VEC coordinating your test session.

The Element 3A and Element 3B question pools were created in March 1987 by dividing the original Element 3 question pool. The subelements were renumbered (subelement 3AA and subelement 3BA came from the original subelement 3A, etc.) but the original question numbers were retained. There were two questions numbered 3A-14.3 and two questions numbered 3F-3.5. One question with each of these numbers is now in the Element 3A pool, and one of each is in the Element 3B pool. The FCC also skipped question 3A-14.4 when they made up the original Element 3 question pool, so there is no question with this number in either the Element 3A or the Element 3B pool.

We have listed page references along with the answers in the answer key section of this chapter. These page numbers indicate where you will find the text discussion related to each question. If you have problems with a question, refer back to the page listed with the answer. You may have to study beyond the listed page numbers. There are no page references for the questions in the Rules and Regulations subelements (3AA and 3BA). Reference material for the rules and regulations questions can be found in *The FCC Rule Book*, published by the ARRL.

One final word about the separation of the Element 3 question pool. While you need to study only the questions in Element 3A if you plan to take an examination for a Technician class license, please remember that you are responsible for questions from *both* Element 3A and Element 3B if you plan to upgrade directly from Novice class (or no license) to General class. If you have a Technician license issued prior to March 21, 1987, you automatically receive credit for both Element 3A and Element 3B, and only need to pass the 13 WPM code test to upgrade to General class.

SUBELEMENT 3BA—Rules and Regulations (4 questions)

3A-3.2   What is the maximum transmitting power permitted an amateur
          station on 10.14-MHz?
          A. 200 watts PEP output
          B. 1000 watts dc input
          C. 1500 watts PEP output
          D. 2000 watts dc input

3A-3.3   What is the maximum transmitting power permitted an amateur
          station on 3725-kHz?
          A. 200 watts PEP output
          B. 1000 watts dc input
          C. 1500 watts PEP output
          D. 2000 watts dc input

3A-3.4   What is the maximum transmitting power permitted an amateur
          station on 7080-kHz?
          A. 200 watts PEP output
          B. 1000 watts dc input
          C. 1500 watts PEP output
          D. 2000 watts dc input

3A-3.5   What is the maximum transmitting power permitted an amateur
          station on 24.95-MHz?
          A. 200 watts PEP output
          B. 1000 watts dc input
          C. 1500 watts PEP output
          D. 2000 watts dc input

3A-3.7   What is the maximum transmitting power permitted an amateur
          station transmitting on 21.150-MHz?
          A. 200 watts PEP output
          B. 1000 watts dc input
          C. 1500 watts dc input
          D. 1500 watts PEP output

3A-4.1   How must a General control operator at a Novice station make the
          station identification when transmitting on 7050-kHz?
          A. The control operator should identify the station with his or her
             call, followed by the word "controlling" and the Novice call
          B. The control operator should identify the station with his or her
             call, followed by the slant bar "/" and the Novice call
          C. The control operator should identify the station with the Novice
             call, followed by the slant bar "/" and his or her own call
          D. A Novice station should not be operated on 7050 kHz, even
             with a General class control operator

3A-4.3   How must a newly-upgraded General control operator with a
          Certificate of Successful Completion of Examination identify the
          station when transmitting on 14.325-MHz pending the receipt of a
          new operator license?
          A. General-class privileges do not include 14.325 MHz
          B. No special form of identification is needed
          C. The operator shall give his/her call sign, followed by the words
             "temporary" and the two-letter ID code shown on the
             certificate of successful completion
          D. The operator shall give his/her call sign, followed by the date
             and location of the VEC examination where he/she obtained
             the upgraded license

3A-6.1　　　Under what circumstances, if any, may third-party traffic be transmitted to a foreign country by an amateur station?
A. Under no circumstances
B. Only if the country has a third-party traffic agreement with the United States
C. Only if the control operator is an Amateur Extra class licensee
D. Only if the country has formal diplomatic relations with the United States

3A-6.2　　　What types of messages may be transmitted by an amateur station to a foreign country for a third-party?
A. Third-party traffic involving material compensation, either tangible or intangible, direct or indirect, to a third party, a station licensee, a control operator, or any other person
B. Third-party traffic consisting of business communications on behalf of any party
C. Only third-party traffic which does not involve material compensation of any kind, and is not business communication of any type
D. No messages may be transmitted to foreign countries for third parties

3A-6.6　　　What additional limitations apply to third-party messages transmitted to foreign countries?
A. Third-party messages may only be transmitted to amateurs in countries with which the US has a third-party traffic agreement
B. Third-party messages may only be sent to amateurs in ITU Region 1
C. Third-party messages may only be sent to amateurs in ITU Region 3
D. Third-party messages must always be transmitted in English

3A-8.6　　　Under what circumstances, if any, may an amateur station transmitting on 29.64-MHz repeat the 146.34-MHz signals of an amateur station with a Technician control operator?
A. Under no circumstances
B. Only if the station on 29.64 MHz is operating under a Special Temporary Authorization allowing such retransmission
C. Only during an FCC-declared general state of communications emergency
D. Only if the control operator of the repeater transmitter is authorized to operate on 29.64 MHz

3A-9.1　　　What frequency privileges are authorized to General operators in the 160 meter band?
A. 1800 to 1900 kHz only
B. 1900 to 2000 kHz only
C. 1800 to 2000 kHz only
D. 1825 to 2000 kHz only

3A-9.2　　　What frequency privileges are authorized to General operators in the 75/80 meter band?
A. 3525 to 3750 and 3850 to 4000 kHz only
B. 3525 to 3775 and 3875 to 4000 kHz only
C. 3525 to 3750 and 3875 to 4000 kHz only
D. 3525 to 3775 and 3850 to 4000 kHz only

3A-9.3   What frequency privileges are authorized to General operators in the 40 meter band?
A. 7025 to 7175 and 7200 to 7300 kHz only
B. 7025 to 7175 and 7225 to 7300 kHz only
C. 7025 to 7150 and 7200 to 7300 kHz only
D. 7025 to 7150 and 7225 to 7300 kHz only

3A-9.4   What frequency privileges are authorized to General operators in the 30 meter band?
A. 10,100 to 10,150 kHz only
B. 10,105 to 10,150 kHz only
C. 10,125 to 10,150 kHz only
D. 10,100 to 10,125 kHz only

3A-9.5   What frequency privileges are authorized to General operators in the 20 meter band?
A. 14,025 to 14,100 and 14,175 to 14,350 kHz only
B. 14,025 to 14,150 and 14,225 to 14,350 kHz only
C. 14,025 to 14,125 and 14,200 to 14,350 kHz only
D. 14,025 to 14,175 and 14,250 to 14,350 kHz only

3A-9.6   What frequency privileges are authorized to General operators in the 15 meter band?
A. 21,025 to 21,200 and 21,275 to 21,450 kHz only
B. 21,025 to 21,150 and 21,300 to 21,450 kHz only
C. 21,025 to 21,200 and 21,300 to 21,450 kHz only
D. 21,000 to 21,150 and 21,275 to 21,450 kHz only

3A-9.7   What frequency privileges are authorized to General operators in the 12 meter band?
A. 24,890 to 24,990 kHz only
B. 24,890 to 24,975 kHz only
C. 24,900 to 24,990 kHz only
D. 24,790 to 24,990 kHz only

3A-9.8   What frequency privileges are authorized to General operators in the 10 meter band?
A. 28,000 to 29,700 kHz only
B. 28,025 to 29,700 kHz only
C. 28,100 to 29,700 kHz only
D. 28,025 to 29,600 kHz only

3A-9.9   Which operator licenses authorize privileges on 1820-kHz?
A. Extra only
B. Extra, Advanced only
C. Extra, Advanced, General only
D. Extra, Advanced, General, Technician only

3A-9.10   Which operator licenses authorize privileges on 3950-kHz?
A. Extra, Advanced only
B. Extra, Advanced, General only
C. Extra, Advanced, General, Technician only
D. Extra, Advanced, General, Technician, Novice only

3A-9.11   Which operator licenses authorize privileges on 7230-kHz?
A. Extra only
B. Extra, Advanced only
C. Extra, Advanced, General only
D. Extra, Advanced, General, Technician only

3A-9.12   Which operator licenses authorize privileges on 10.125-MHz?
A. Extra, Advanced, General only
B. Extra, Advanced only
C. Extra only
D. Technician only

3A-9.13   Which operator licenses authorize privileges on 14.325-MHz?
A. Extra, Advanced, General, Technician only
B. Extra, Advanced, General only
C. Extra, Advanced only
D. Extra only

3A-9.14   Which operator licenses authorize privileges on 21.425-MHz?
A. Extra, Advanced, General, Novice only
B. Extra, Advanced, General, Technician only
C. Extra, Advanced, General only
D. Extra, Advanced only

3A-9.15   Which operator licenses authorize privileges on 24.895-MHz?
A. Extra only
B. Extra, Advanced only
C. Extra, Advanced, General only
D. None

3A-9.16   Which operator licenses authorize privileges on 29.616-MHz?
A. Novice, Technician, General, Advanced, Extra only
B. Technician, General, Advanced, Extra only
C. General, Advanced, Extra only
D. Advanced, Extra only

3A-10.1   On what frequencies within the 160 meter band may emission A3E
be transmitted?
A. 1800-2000 kHz only
B. 1800-1900 kHz only
C. 1900-2000 kHz only
D. 1825-1950 kHz only

3A-10.2   On what frequencies within the 80 meter band may emission A1A
be transmitted?
A. 3500-3750 kHz only
B. 3700-3750 kHz only
C. 3500-4000 kHz only
D. 3890-4000 kHz only

3A-10.3   On what frequencies within the 40 meter band may emission A3F
be transmitted?
A. 7225-7300 kHz only
B. 7000-7300 kHz only
C. 7100-7150 kHz only
D. 7150-7300 kHz only

3A-10.4   On what frequencies within the 30 meter band may emission F1B
be transmitted?
A. 10.140-10.150 MHz only
B. 10.125-10.150 MHz only
C. 10.100-10.150 MHz only
D. 10.100-10.125 MHz only

3A-10.5   On what frequencies within the 20 meter band may emission A3C
           be transmitted?
           A. 14,200-14,300 kHz only
           B. 14,150-14,350 kHz only
           C. 14,025-14,150 kHz only
           D. 14,150-14,300 kHz only

3A-10.6   On what frequencies within the 15 meter band may emission F3C
           be transmitted?
           A. 21,200-21,300 kHz only
           B. 21,350-21,450 kHz only
           C. 21,200-21,450 kHz only
           D. 21,100-21,200 kHz only

3A-10.7   On what frequencies within the 12 meter band may emission J3E be
           transmitted?
           A. 24,890-24,990 kHz only
           B. 24,890-24,930 kHz only
           C. 24,930-24,990 kHz only
           D. J3E is not permitted in this band

3A-10.8   On what frequencies within the 10 meter band may emission A3E
           be transmitted?
           A. 28,000-28,300 kHz only
           B. 29,000-29,700 kHz only
           C. 28,300-29,700 kHz only
           D. 28,000-29,000 kHz only

3A-13.1   How is the *sending speed* (signaling rate) for digital communications
           determined?
           A. By taking the reciprocal of the shortest (signaling) time interval
              (in minutes) that occurs during a transmission, where each
              time interval is the period between changes of transmitter state
              (including changes in emission amplitude, frequency, phase, or
              combination of these, as authorized)
           B. By taking the square root of the shortest (signaling) time
              interval (in seconds) that occurs during a transmission, where
              each time interval is the period between changes of transmitter
              state (including changes in emission amplitude, frequency,
              phase, or combination of these, as authorized)
           C. By taking the reciprocal of the shortest (signaling) time interval
              (in seconds) that occurs during a transmission, where each
              time interval is the period between changes of transmitter
              state (including changes in emission amplitude, frequency,
              phase, or combination of these, as authorized)
           D. By taking the square root of the shortest (signaling) time
              interval (in minutes) that occurs during a transmission, where
              each time interval is the period between changes of
              transmitter state (including changes in emission amplitude,
              frequency, phase, or combination of these, as authorized)

3A-13.2   What is the maximum sending speed permitted for an emission F1B
           transmission below 28-MHz?
           A. 56 kilobauds
           B. 19.6 kilobauds
           C. 1200 bauds
           D. 300 bauds

3A-14.3   Under what circumstances, if any, may an amateur station engage in some form of broadcasting?
- A. During severe storms, amateurs may broadcast weather information for people with scanners
- B. Under no circumstances
- C. If power levels under one watt are used, amateur stations may broadcast information bulletins, but not music
- D. Amateur broadcasting is permissible above 10 GHz

**3A-14.4   THIS QUESTION WAS SKIPPED WHEN FCC MADE UP THE QUESTION POOL. THERE IS NO QUESTION WITH THIS NUMBER.

3A-14.6   What protection, if any, is afforded an amateur station transmission against retransmission by a broadcast station?
- A. No protection whatsoever
- B. The broadcaster must secure permission for retransmission from the control operator of the amateur station
- C. The broadcaster must petition the FCC for retransmission rights 30 days in advance
- D. Retransmissions may only be made during a declared emergency

3A-15.1   Under what circumstances, if any, may the playing of a violin be transmitted by an amateur station?
- A. When the music played produces no dissonances or spurious emissions
- B. When it is used to jam an illegal transmission
- C. Only above 1215 MHz
- D. Transmitting music is not permitted in the Amateur Service

3A-15.3   Under what circumstances, if any, may the playing of a piano be transmitted by an amateur station?
- A. When it is used to jam an illegal transmission
- B. Only above 1215 MHz
- C. Transmitting music is not permitted in the Amateur Service
- D. When the music played produces no dissonances or spurious emissions

3A-15.4   Under what circumstances, if any, may the playing of a harmonica be transmitted by an amateur station?
- A. When the music played produces no dissonances or spurious emissions
- B. Transmitting music is not permitted in the Amateur Service
- C. When it is used to jam an illegal transmission
- D. Only above 1215 MHz

3A-16.1   Under what circumstances, if any, may an amateur station transmit a message in a secret code in order to obscure the meaning?
- A. Only above 450 MHz
- B. Only on Field Day
- C. Never
- D. Only during a declared communications emergency

3A-16.2   What types of abbreviations or signals are not considered codes or ciphers?
   A. Abbreviations and signals certified by the ARRL
   B. Abbreviations and signals established by regulation or custom and usage and whose intent is to facilitate communication and not to obscure meaning
   C. No abbreviations are permitted, as they tend to obscure the meaning of the message to FCC monitoring stations
   D. Only "10-codes" are permitted

3A-16.3   When, if ever, are codes and ciphers permitted in domestic amateur radiocommunications?
   A. Codes and ciphers are prohibited under all circumstances
   B. Codes and ciphers are permitted during ARRL-sponsored contests
   C. Codes and ciphers are permitted during nationally declared emergencies
   D. Codes and ciphers are permitted above 2.3 GHz

3A-16.4   When, if ever, are codes and ciphers permitted in international amateur radiocommunications?
   A. Codes and ciphers are prohibited under all circumstances
   B. Codes and ciphers are permitted during ITU-sponsored DX contests
   C. Codes and ciphers are permitted during internationally declared emergencies
   D. Codes and ciphers are permitted only on frequencies above 2.3 GHz

3B-1.4   What is meant by the term *flattopping* in an emission J3E
         transmission?
         A. Signal distortion caused by insufficient collector current
         B. The transmitter's automatic level control is properly adjusted
         C. Signal distortion caused by excessive drive
         D. The transmitter's carrier is properly suppressed

3B-1.5   How should the microphone gain control be adjusted on an
         emission J3E transmitter?
         A. For full deflection of the ALC meter on modulation peaks
         B. For slight movement of the ALC meter on modulation peaks
         C. For 100% frequency deviation on modulation peaks
         D. For a dip in plate current

3B-2.1   In which segment of the 20 meter band do most emission F1B
         transmissions take place?
         A. Between 14.000 and 14.050 MHz
         B. Between 14.075 and 14.100 MHz
         C. Between 14.150 and 14.225 MHz
         D. Between 14.275 and 14.350 MHz

3B-2.2   In which segment of the 80 meter band do most emission F1B
         transmissions take place?
         A. 3.610 to 3.630 MHz
         B. 3500 to 3525 kHz
         C. 3700 to 3750 kHz
         D. 3.775 to 3.825 MHz

3B-2.3   What is meant by the term *Baudot*?
         A. Baudot is a 7-bit code, with start, stop and parity bits
         B. Baudot is a 7-bit code in which each character has four mark
            and three space bits
         C. Baudot is a 5-bit code, with additional start and stop bits
         D. Baudot is a 6-bit code, with additional start, stop and parity
            bits

3B-2.4   What is meant by the term *ASCII*?
         A. ASCII is a 7-bit code, with additional start, stop and parity bits
         B. ASCII is a 7-bit code in which each character has four mark
            and three space bits
         C. ASCII is a 5-bit code, with additional start and stop bits
         D. ASCII is a 5-bit code in which each character has three mark
            and two space bits

3B-2.6   What is the most common frequency shift for emission F1B
         transmissions in the amateur HF bands?
         A. 85 Hz
         B. 170 Hz
         C. 425 Hz
         D. 850 Hz

3B-2.10 What are the two subset modes of AMTOR?
         A. A mark of 2125 Hz and a space of 2295 Hz
         B. Baudot and ASCII
         C. ARQ and FEC
         D. USB and LSB

3B-2.11   What is the meaning of the term *ARQ*?
            A. Automatic Repeater Queue
            B. Automatic Receiver Quieting
            C. Automatically Resend Quickly
            D. Automatic Repeat Request

3B-2.12   What is the meaning of the term *FEC*?
            A. Frame Error Check
            B. Forward Error Correction
            C. Frequency Envelope Control
            D. Frequency Encoded Connection

3B-3.8   What is a *band plan*?
            A. An outline adopted by Amateur Radio operators for operating
               within a specific portion of radio spectrum
            B. An arrangement for deviating from FCC Rules and Regulations
            C. A schedule for operating devised by the Federal
               Communications Commission
            D. A plan devised for a club on how best to use a band during a
               contest

3B-3.12   What is the usual input/output frequency separation for a 10 meter
            station in repeater operation?
            A. 100 kHz
            B. 600 kHz
            C. 1.6 MHz
            D. 170 Hz

3B-4.1   What is meant by the term *VOX transmitter control*?
            A. Circuitry that causes the transmitter to transmit automatically
               when the operator speaks into the microphone
            B. Circuitry that shifts the frequency of the transmitter when the
               operator switches from radiotelegraphy to radiotelephony
            C. Circuitry that activates the receiver incremental tuning in a
               transceiver
            D. Circuitry that isolates the microphone from the ambient noise
               level

3B-4.2   What is the common name for the circuit that causes a transmitter
            to automatically transmit when a person speaks into the
            microphone?
            A. VXO
            B. VOX
            C. VCO
            D. VFO

3B-5.1   What is meant by the term *full break-in telegraphy*?
            A. A system of radiotelegraph communication in which the
               breaking station sends the Morse Code symbols BK
            B. A system of radiotelegraph communication in which only
               automatic keyers can be used
            C. A system of radiotelegraph communication in which the
               operator must activate the send-receive switch after
               completing a transmission
            D. A system of radiotelegraph communication in which the
               receiver is sensitive to incoming signals between transmitted
               key pulses

3B-5.2   What Q signal is used to indicate full break-in telegraphy capability?
          A. QSB
          B. QSF
          C. QSK
          D. QSV

3B-6.1   When selecting an emission A1A transmitting frequency, what is the
          minimum frequency separation from a QSO in progress that should
          be allowed in order to minimize interference?
          A. 5 to 50 Hz
          B. 150 to 500 Hz
          C. Approximately 3 kHz
          D. Approximately 6 kHz

3B-6.2   When selecting an emission J3E transmitting frequency, what is the
          minimum frequency separation from a QSO in progress that should
          be allowed in order to minimize interference?
          A. 150 to 500 Hz between suppressed carriers
          B. Approximately 3 kHz between suppressed carriers
          C. Approximately 6 kHz between suppressed carriers
          D. Approximately 10 kHz between suppressed carriers

3B-6.3   When selecting an emission F1B RTTY transmitting frequency, what
          is the minimum frequency separation from a QSO in progress that
          should be allowed in order to minimize interference?
          A. Approximately 45 Hz center to center
          B. Approximately 250 to 500 Hz center to center
          C. Approximately 3 kHz center to center
          D. Approximately 6 kHz center to center

3B-7.1   What is an *azimuthal* map?
          A. A map projection that is always centered on the North Pole
          B. A map projection, centered on a particular location, that
             determines the shortest path between two points on the
             surface of the earth
          C. A map that shows the angle at which an amateur satellite
             crosses the equator
          D. A map that shows the number of degrees longitude that an
             amateur satellite appears to move westward at the equator
             with each orbit

3B-7.2   How can an azimuthal map be helpful in conducting international
          HF radiocommunications?
          A. It is used to determine the proper beam heading for the
             shortest path to a DX station
          B. It is used to determine the most efficient transmitting antenna
             height to conduct the desired communication
          C. It is used to determine the angle at which an amateur satellite
             crosses the equator
          D. It is used to determine the maximum usable frequency (MUF)

3B-7.3   What is the most useful type of map when orienting a directional
          antenna toward a station 5,000 miles distant?
          A. Azimuthal
          B. Mercator
          C. Polar projection
          D. Topographical

3B-7.4   A directional antenna pointed in the long-path direction to another
          station is generally oriented how many degrees from the short-path
          heading?
          A. 45 degrees
          B. 90 degrees
          C. 180 degrees
          D. 270 degrees

3B-7.5   What is the short-path heading to Antarctica?
          A. Approximately 0 degrees
          B. Approximately 90 degrees
          C. Approximately 180 degrees
          D. Approximately 270 degrees

3B-8.1   When permitted, transmissions to amateur stations in another
          country must be limited to only what type of messages?
          A. Messages of any type are permitted
          B. Messages that compete with public telecommunications
              services
          C. Messages of a technical nature or remarks of a personal
              character of relative unimportance
          D. Such transmissions are never permitted

3B-8.2   In which International Telecommunication Union Region is the
          continental United States?
          A. Region 1
          B. Region 2
          C. Region 3
          D. Region 4

3B-8.3   In which International Telecommunication Union Region is Alaska?
          A. Region 1
          B. Region 2
          C. Region 3
          D. Region 4

3B-8.4   In which International Telecommunication Union Region is American
          Samoa?
          A. Region 1
          B. Region 2
          C. Region 3
          D. Region 4

3B-8.5   For uniformity in international radiocommunication, what time
          measurement standard should amateur radio operators worldwide
          use?
          A. Eastern Standard Time
          B. Uniform Calibrated Time
          C. Coordinated Universal Time
          D. Universal Time Control

3B-8.6   In which International Telecommunication Union Region is Hawaii?
          A. Region 1
          B. Region 2
          C. Region 3
          D. Region 4

3B-8.7 In which International Telecommunication Union Region are the Northern Mariana Islands?
    A. Region 1
    B. Region 2
    C. Region 3
    D. Region 4

3B-8.8 In which International Telecommunication Union Region is Guam?
    A. Region 1
    B. Region 2
    C. Region 3
    D. Region 4

3B-8.9 In which International Telecommunication Union Region is Wake Island?
    A. Region 1
    B. Region 2
    C. Region 3
    D. Region 4

3B-10.1 What is the *Amateur Auxiliary* to the FCC's Field Operations Bureau?
    A. Amateur Volunteers formally enlisted to monitor the airwaves for rules violations
    B. Amateur Volunteers who conduct Amateur Radio licensing examinations
    C. Amateur Volunteers who conduct frequency coordination for amateur VHF repeaters
    D. Amateur Volunteers who determine height above average terrain measurements for repeater installations

3B-10.2 What are the objectives of the Amateur Auxiliary to the FCC's Field Operations Bureau?
    A. To enforce amateur self-regulation and compliance with the rules
    B. To foster amateur self-regulation and compliance with the rules
    C. To promote efficient and orderly spectrum usage in the repeater subbands
    D. To provide emergency and public safety communications

SUBELEMENT 3BC—Radio-Wave Propagation (3 questions)

3C-1.6   What is the maximum distance along the earth's surface that can normally be covered in one hop using the F2 layer?
A. Approximately 180 miles
B. Approximately 1200 miles
C. Approximately 2500 miles
D. No distance. This layer does not support radio communication

3C-1.7   What is the maximum distance along the earth's surface that can be covered in one hop using the E layer?
A. Approximately 180 miles
B. Approximately 1200 miles
C. Approximately 2500 miles
D. No distance. This layer does not support radio communication

3C-1.9   What is the average height of maximum ionization of the E layer?
A. 45 miles
B. 70 miles
C. 200 miles
D. 1200 miles

3C-1.10   During what part of the day, and in what season of the year can the F2 layer be expected to reach its maximum height?
A. At noon during the summer
B. At midnight during the summer
C. At dusk in the spring and fall
D. At noon during the winter

3C-1.13   What is the *critical angle*, as used in radio wave propagation?
A. The lowest take off angle that will return a radio wave to earth under specific ionospheric conditions
B. The compass direction of the desired DX station from your location
C. The 180-degree-inverted compass direction of the desired DX station from your location
D. The highest take off angle that will return a radio wave to earth during specific ionospheric conditions

3C-2.3   What is the main reason that the 160, 80, and 40 meter amateur bands tend to be useful for only short-distance communications during daylight hours?
A. Because of a lack of activity
B. Because of auroral propagation
C. Because of D-layer absorption
D. Because of magnetic flux

3C-2.4   What is the principal reason the 160 meter through 40 meter bands are useful for only short-distance radiocommunications during daylight hours?
A. F-layer bending
B. Gamma radiation
C. D-layer absorption
D. Tropospheric ducting

3C-3.3   If the maximum usable frequency on the path from Minnesota to Africa is 22-MHz, which band should offer the best chance for a successful QSO?
   A. 10 meters
   B. 15 meters
   C. 20 meters
   D. 40 meters

3C-3.4   If the maximum usable frequency on the path from Ohio to West Germany is 17-MHz, which band should offer the best chance for a successful QSO?
   A. 80 meters
   B. 40 meters
   C. 20 meters
   D. 2 meters

3C-5.1   Over what periods of time do sudden ionospheric disturbances normally last?
   A. The entire day
   B. A few minutes to a few hours
   C. A few hours to a few days
   D. Approximately one week

3C-5.2   What can be done at an amateur station to continue radio-communications during a sudden ionospheric disturbance?
   A. Try a higher frequency
   B. Try the other sideband
   C. Try a different antenna polarization
   D. Try a different frequency shift

3C-5.3   What effect does a sudden ionospheric disturbance have on the daylight ionospheric propagation of HF radio waves?
   A. Disrupts higher-latitude paths more than lower-latitude paths
   B. Disrupts transmissions on lower frequencies more than those on higher frequencies
   C. Disrupts communications via satellite more than direct communications
   D. None. Only dark (as in nighttime) areas of the globe are affected

3C-5.4   How long does it take a solar disturbance that increases the sun's ultraviolet radiation to cause ionospheric disturbances on earth?
   A. Instantaneously
   B. 1.5 seconds
   C. 8 minutes
   D. 20 to 40 hours

3C-5.5   Sudden ionospheric disturbances cause increased radio wave absorption in which layer of the ionosphere?
   A. D layer
   B. E layer
   C. F1 layer
   D. F2 layer

3C-6.2   What is a characteristic of *backscatter* signals?
   A. High intelligibility
   B. A wavering sound
   C. Reversed modulation
   D. Reversed sidebands

3C-6.4   What makes backscatter signals often sound distorted?
           A. Auroral activity and changes in the earth's magnetic field
           B. The propagation through ground waves that absorb much of
              the signal's clarity
           C. The earth's E-layer at the point of radio wave refraction
           D. The small part of the signal's energy scattered back to the
              transmitter skip zone through several radio-wave paths

3C-6.5   What is the radio wave propagation phenomenon that allows a
           signal to be detected at a distance too far for ground wave
           propagation but too near for normal sky wave propagation?
           A. Ground wave
           B. Scatter
           C. Sporadic-E skip
           D. Short path skip

3C-6.6   When does ionospheric scatter propagation on the HF bands most
           often occur?
           A. When the sunspot cycle is at a minimum
           B. At night
           C. When the F1 and F2 layers are combined
           D. At frequencies above the maximum usable frequency

3C-7.1   What is *solar flux*?
           A. The density of the sun's magnetic field
           B. The radio energy emitted by the sun
           C. The number of sunspots on the side of the sun facing the
              earth
           D. A measure of the tilt of the earth's ionosphere on the side
              toward the sun

3C-7.2   What is the *solar-flux index*?
           A. A measure of past measurements of solar activity
           B. A measurement of solar activity that compares daily readings
              with results from the last six months
           C. Another name for the American sunspot number
           D. A measure of solar activity that is taken daily

3C-7.3   What is a timely indicator of solar activity?
           A. The 2800-MHz solar flux index
           B. The mean Canadian sunspot number
           C. A clock set to Coordinated Universal Time
           D. Van Allen radiation measurements taken at Boulder, Colorado

3C-7.4   What type of propagation conditions on the 15 meter band are
           indicated by a solar-flux index value of 60 to 70?
           A. Unpredictable ionospheric propagation
           B. No ionospheric propagation is possible
           C. Excellent ionospheric propagation
           D. Poor ionospheric propagation

3C-7.5   A solar flux index in the range of 90 to 110 indicates what type of
           propagation conditions on the 15 meter band?
           A. Poor ionospheric propagation
           B. No ionospheric propagation is possible
           C. Unpredictable ionospheric propagation
           D. Good ionospheric propagation

3C-7.6   A solar flux index of greater than 120 would indicate what type of propagation conditions on the 10 meter band?
A. Good ionospheric propagation
B. Poor ionospheric propagation
C. No ionospheric propagation is possible
D. Unpredictable ionospheric propagation

3C-7.7   For widespread long distance openings on the 6 meter band, what solar-flux index values would be required?
A. Less than 50
B. Approximately 75
C. Greater than 100
D. Greater than 250

3C-7.8   If the MUF is high and HF radiocommunications are generally good for several days, a similar condition can usually be expected how many days later?
A. 7 days
B. 14 days
C. 28 days
D. 90 days

3C-10.1   What is a *geomagnetic disturbance*?
A. A sudden drop in the solar-flux index
B. A shifting of the earth's magnetic pole
C. Ripples in the ionosphere
D. A dramatic change in the earth's magnetic field over a short period of time

3C-10.2   Which latitude paths are more susceptible to geomagnetic disturbances?
A. Those greater than 45 degrees latitude
B. Those less than 45 degrees latitude
C. Equatorial paths
D. All paths are affected equally

3C-10.3   What can be the effect of a major geomagnetic storm on radiocommunications?
A. Improved high-latitude HF communications
B. Degraded high-latitude HF communications
C. Improved ground-wave propagation
D. Improved chances of ducting at UHF

3C-10.4   How long does it take a solar disturbance that increases the sun's radiation of charged particles to affect radio wave propagation on earth?
A. The effect is instantaneous
B. 1.5 seconds
C. 8 minutes
D. 20 to 40 hours

3D-1.5   Which wires in a four conductor line cord should be attached to
         fuses in a 234-vac primary (single phase) power supply?
         A. Only the "hot" (black and red) wires
         B. Only the "neutral" (white) wire
         C. Only the ground (bare) wire
         D. All wires

3D-1.6   What size wire is normally used on a 15-ampere, 117-vac household
         lighting circuit?
         A. AWG number 14
         B. AWG number 16
         C. AWG number 18
         D. AWG number 22

3D-1.7   What size wire is normally used on a 20-ampere, 117-vac household
         appliance circuit?
         A. AWG number 20
         B. AWG number 16
         C. AWG number 14
         D. AWG number 12

3D-1.8   What could be a cause of the room lights dimming when the
         transmitter is keyed?
         A. RF in the ac pole transformer
         B. High resistance in the key contacts
         C. A drop in ac line voltage
         D. The line cord is wired incorrectly

3D-1.9   What size fuse should be used on a #12 wire household appliance
         circuit?
         A. Maximum of 100 amperes
         B. Maximum of 60 amperes
         C. Maximum of 30 amperes
         D. Maximum of 20 amperes

3D-2.4   What safety feature is provided by a bleeder resistor in a power
         supply?
         A. It improves voltage regulation
         B. It discharges the filter capacitors
         C. It removes shock hazards from the induction coils
         D. It eliminates ground-loop current

3D-3.1   What kind of input signal is used to test the amplitude linearity of an
         emission J3E transmitter while viewing the output on an
         oscilloscope?
         A. Normal speech
         B. An audio-frequency sine wave
         C. Two audio-frequency sine waves
         D. An audio-frequency square wave

3D-3.2   To test the amplitude linearity of an emission J3E transmitter with
         an oscilloscope, what should the audio input to the transmitter be?
         A. Normal speech
         B. An audio-frequency sine wave
         C. Two audio-frequency sine waves
         D. An audio-frequency square wave

3D-3.3   How are two-tones used to test the amplitude linearity of an
          emission J3E transmitter?
          A. Two harmonically related audio tones are fed into the
             microphone input of a J3E transmitter, and the output is
             observed on an oscilloscope
          B. Two harmonically related audio tones are fed into the
             microphone input of the transmitter, and the output is
             observed on a distortion analyzer
          C. Two non-harmonically related audio tones are fed into the
             microphone input of the transmitter, and the output is
             observed on an oscilloscope
          D. Two non-harmonically related audio tones are fed into the
             microphone input of the transmitter, and the output is
             observed on a wattmeter

3D-3.4   What audio frequencies are used in a *two-tone test* of the linearity
          of an emission J3E transmitter?
          A. 20 Hz and 20,000 Hz tones must be used
          B. 1200 Hz and 2400 Hz tones must be used
          C. Any two audio tones may be used, if they are harmonically
             related
          D. Any two audio tones may be used, but they must be within the
             transmitter audio passband, and should not be harmonically
             related

3D-3.5   What can be determined by making a *two-tone test* using an
          oscilloscope?
          A. The percent of frequency modulation
          B. The percent of carrier phase shift
          C. The frequency deviation
          D. The amplifier linearity

3D-4.1   How can the grid-current meter in a power amplifier be used as a
          neutralizing indicator?
          A. Tune for minimum change in grid current as the output circuit
             is changed
          B. Tune for maximum change in grid current as the output circuit
             is changed
          C. Tune for minimum grid current
          D. Tune for maximum grid current

3D-4.2   Why is neutralization in some vacuum tube amplifiers necessary?
          A. To reduce the limits of loaded Q in practical tuned circuits
          B. To reduce grid to cathode leakage
          C. To cancel acid build-up caused by thorium oxide gas
          D. To cancel oscillation caused by the effects of interelectrode
             capacitance

3D-4.3   How is neutralization of an rf amplifier accomplished?
          A. By supplying energy from the amplifier output to the input on
             alternate half cycles
          B. By supplying energy from the amplifier output to the input
             shifted 360 degrees out of phase
          C. By supplying energy from the amplifier output to the input
             shifted 180 degrees out of phase
          D. By supplying energy from the amplifier output to the input with
             a proper dc bias

3D-4.4 What purpose does a neutralizing circuit serve in an rf amplifier?
A. It controls differential gain
B. It cancels the effects of positive feedback
C. It eliminates circulating currents
D. It reduces incidental grid modulation

3D-4.5 What is the reason for neutralizing the final amplifier stage of a transmitter?
A. To limit the modulation index
B. To eliminate parasitic oscillations
C. To cut off the final amplifier during standby periods
D. To keep the carrier on frequency

3D-5.1 How can the output PEP of a transmitter be determined with an oscilloscope?
A. Measure peak load voltage across a resistive load with an oscilloscope, and calculate, using PEP = [(Vp)(Vp)] / (RL)
B. Measure peak load voltage across a resistive load with an oscilloscope, and calculate, using PEP = [(0.707 PEV) (0.707 PEV)] / RL
C. Measure peak load voltage across a resistive load with an oscilloscope, and calculate, using PEP = (Vp)(Vp)(RL)
D. Measure peak load voltage across a resistive load with an oscilloscope, and calculate, using PEP = [(1.414 PEV) (1.414 PEV)] / RL

3D-5.5 What is the output PEP from a transmitter when an oscilloscope shows 200-volts peak-to-peak across a 50 ohm resistor connected to the transmitter output terminals?
A. 100 watts
B. 200 watts
C. 400 watts
D. 1000 watts

3D-5.6 What is the output PEP from a transmitter when an oscilloscope shows 500-volts peak-to-peak across a 50 ohm resistor connected to the transmitter output terminals?
A. 500 watts
B. 625 watts
C. 1250 watts
D. 2500 watts

3D-5.7 What is the output PEP from an NØN transmitter when an average-reading wattmeter connected to the transmitter output terminals indicates 1060 watts?
A. 530 watts
B. 1060 watts
C. 1500 watts
D. 2120 watts

3D-6.1 What item of test equipment contains horizontal and vertical channel amplifiers?
A. The ohmmeter
B. The signal generator
C. The ammeter
D. The oscilloscope

3D-6.2   What types of signals can an oscilloscope measure?
- A. Any time-dependent signal within the bandwidth capability of the instrument
- B. Blinker-light signals from ocean-going vessels
- C. International nautical flag signals
- D. Signals created by aeronautical flares

3D-6.3   What is an *oscilloscope*?
- A. An instrument that displays the radiation resistance of an antenna
- B. An instrument that displays the SWR on a feed line
- C. An instrument that displays the resistance in a circuit
- D. An instrument that displays signal waveforms

3D-6.4   What can cause phosphor damage to an oscilloscope cathode ray tube?
- A. Directly connecting deflection electrodes to the cathode ray tube
- B. Too high an intensity setting
- C. Overdriving the vertical amplifier
- D. Improperly adjusted focus

3D-9.1   What is a *signal tracer*?
- A. A direction-finding antenna
- B. An aid for following schematic diagrams
- C. A device for detecting signals in a circuit
- D. A device for drawing signal waveforms

3D-9.2   How is a signal tracer used?
- A. To detect the presence of a signal in the various stages of a receiver
- B. To locate a source of interference
- C. To trace the path of a radio signal through the ionosphere
- D. To draw a waveform on paper

3D-9.3   What is a signal tracer normally used for?
- A. To identify the source of radio transmissions
- B. To make exact replicas of signals
- C. To give a visual indication of standing waves on open-wire feed lines
- D. To identify an inoperative stage in a radio receiver

3D-10.1   What is the most effective way to reduce or eliminate audio frequency interference to home entertainment systems?
- A. Install bypass inductors
- B. Install bypass capacitors
- C. Install metal oxide varistors
- D. Install bypass resistors

3D-10.2   What should be done when a properly-operating amateur station is the source of interference to a nearby telephone?
- A. Make internal adjustments to the telephone equipment
- B. Contact a phone service representative about installing RFI filters
- C. Nothing can be done to cure the interference
- D. Ground and shield the local telephone distribution amplifier

3D-10.3   What sound is heard from a public address system when audio rectification occurs in response to a nearby emission J3E transmission?
   A. A steady hum that persists while the transmitter's carrier is on the air
   B. On-and-off humming or clicking
   C. Distorted speech from the transmitter's signals
   D. Clearly audible speech from the transmitter's signals

3D-10.4   How can the possibility of audio rectification occurring be minimized?
   A. By using a solid state transmitter
   B. By using CW emission only
   C. By ensuring all station equipment is properly grounded
   D. By using AM emission only

3D-10.5   What sound is heard from a public address system when audio rectification occurs in response to a nearby emission A3E transmission?
   A. Audible, possibly distorted speech from the transmitter signals
   B. On-and-off humming or clicking
   C. Muffled, distorted speech from the transmitter's signals
   D. Extremely loud, severely distorted speech from the transmitter's signals

3D-12.2   What is the reason for using a speech processor with an emission J3E transmitter?
   A. A properly adjusted speech processor reduces average transmitter power requirements
   B. A properly adjusted speech processor reduces unwanted noise pickup from the microphone
   C. A properly adjusted speech processor improves voice frequency fidelity
   D. A properly adjusted speech processor improves signal intelligibility at the receiver

3D-12.3   When a transmitter is 100% modulated, will a speech processor increase the output PEP?
   A. Yes
   B. No
   C. It will decrease the transmitter's peak power output
   D. It will decrease the transmitter's average power output

3D-12.4   Under which band conditions should a speech processor not be used?
   A. When there is high atmospheric noise on the band
   B. When the band is crowded
   C. When the frequency in use is clear
   D. When the sunspot count is relatively high

3D-12.5   What effect can result from using a speech processor with an emission J3E transmitter?
   A. A properly adjusted speech processor reduces average transmitter power requirements
   B. A properly adjusted speech processor reduces unwanted noise pickup from the microphone
   C. A properly adjusted speech processor improves voice frequency fidelity
   D. A properly adjusted speech processor improves signal intelligibility at the receiver

3D-13.1   At what point in a coaxial line should an electronic T-R switch be installed?
A. Between the transmitter and low-pass filter
B. Between the low-pass filter and antenna
C. At the antenna feed point
D. Right after the low-pass filter

3D-13.2   Why is an electronic T-R switch preferable to a mechanical one?
A. Greater receiver sensitivity
B. Circuit simplicity
C. Higher operating speed
D. Cleaner output signals

3D-13.3   What station accessory facilitates QSK operation?
A. Oscilloscope
B. Audio CW filter
C. Antenna relay
D. Electronic TR switch

3D-14.6   What is an antenna *noise bridge*?
A. An instrument for measuring the noise figure of an antenna or other electrical circuit
B. An instrument for measuring the impedance of an antenna or other electrical circuit
C. An instrument for measuring solar flux
D. An instrument for tuning out noise in a receiver

3D-14.7   How is an antenna noise bridge used?
A. It is connected at the antenna feed point, and the noise is read directly
B. It is connected between a transmitter and an antenna and tuned for minimum SWR
C. It is connected between a receiver and an unknown impedance and tuned for minimum noise
D. It is connected between an antenna and a Transmatch and adjusted for minimum SWR

3D-15.1   How does the emitted waveform from a properly-adjusted emission J3E transmitter appear on a monitoring oscilloscope?
A. A vertical line
B. A waveform that mirrors the input waveform
C. A square wave
D. Two loops at right angles

3D-15.2   What is the best instrument for checking transmitted signal quality from an emissions A1A/J3E transmitter?
A. A monitor oscilloscope
B. A field strength meter
C. A sidetone monitor
D. A diode probe and an audio amplifier

3D-15.3   What is a *monitoring oscilloscope*?
A. A device used by the FCC to detect out-of-band signals
B. A device used to observe the waveform of a transmitted signal
C. A device used to display SSTV signals
D. A device used to display signals in a receiver IF stage

3D-15.4   How is a monitoring oscilloscope connected in a station in order to
           check the quality of the transmitted signal?
           A. Connect the receiver IF output to the vertical-deflection plates
              of the oscilloscope
           B. Connect the transmitter audio input to the oscilloscope vertical
              input
           C. Connect a receiving antenna directly to the oscilloscope
              vertical input
           D. Connect the transmitter output to the vertical-deflection plates
              of the oscilloscope

3D-17.2   What is the most appropriate instrument to use when determining
           antenna horizontal radiation patterns?
           A. A field strength meter
           B. A grid-dip meter
           C. A wave meter
           D. A vacuum-tube voltmeter

3D-17.3   What is a *field-strength* meter?
           A. A device for determining the standing-wave ratio on a
              transmission line
           B. A device for checking modulation on the output of a
              transmitter
           C. A device for monitoring relative RF output
           D. A device for increasing the average transmitter output

3D-17.4   What is a simple instrument that can be useful for monitoring
           relative rf output during antenna and transmitter adjustments?
           A. A field-strength meter
           B. An antenna noise bridge
           C. A multimeter
           D. A Transmatch

3D-17.5   When the power output from a transmitter is increased by four
           times, how should the S-meter reading on a nearby receiver
           change?
           A. Decrease by approximately one S-unit
           B. Increase by approximately one S-unit
           C. Increase by approximately four S-units
           D. Decrease by approximately four S-units

3D-17.6   By how many times must the power output from a transmitter be
           increased to raise the S-meter reading on a nearby receiver from
           S-8 to S-9?
           A. Approximately 2 times
           B. Approximately 3 times
           C. Approximately 4 times
           D. Approximately 5 times

SUBELEMENT 3BE—Electrical Principles (2 questions)

3E-1.1   What is meant by the term *impedance*?
          A. The electric charge stored by a capacitor
          B. The opposition to the flow of ac in a circuit containing only
             capacitance
          C. The opposition to the flow of ac in a circuit
          D. The force of repulsion presented to an electric field by another
             field with the same charge

3E-1.2   What is the opposition to the flow of ac in a circuit containing both
          resistance and reactance called?
          A. Ohm
          B. Joule
          C. Impedance
          D. Watt

3E-3.1   What is meant by the term *reactance*?
          A. Opposition to dc caused by resistors
          B. Opposition to ac caused by inductors and capacitors
          C. A property of ideal resistors in ac circuits
          D. A large spark produced at switch contacts when an inductor is
             de-energized

3E-3.2   What is the opposition to the flow of ac caused by an inductor
          called?
          A. Resistance
          B. Reluctance
          C. Admittance
          D. Reactance

3E-3.3   What is the opposition to the flow of ac caused by a capacitor
          called?
          A. Resistance
          B. Reluctance
          C. Admittance
          D. Reactance

3E-3.4   How does a coil react to ac?
          A. As the frequency of the applied ac increases, the reactance
             decreases
          B. As the amplitude of the applied ac increases, the reactance
             also increases
          C. As the amplitude of the applied ac increases, the reactance
             decreases
          D. As the frequency of the applied ac increases, the reactance
             also increases

3E-3.5   How does a capacitor react to ac?
          A. As the frequency of the applied ac increases, the reactance
             decreases
          B. As the frequency of the applied ac increases, the reactance
             increases
          C. As the amplitude of the applied ac increases, the reactance
             also increases
          D. As the amplitude of the applied ac increases, the reactance
             decreases

3E-6.1   When will a power source deliver maximum output?
          A. When the impedance of the load is equal to the impedance of
             the source
          B. When the SWR has reached a maximum value
          C. When the power supply fuse rating equals the primary winding
             current
          D. When air wound transformers are used instead of iron core
             transformers

3E-6.2   What is meant by *impedance matching*?
          A. To make the load impedance much greater than the source
             impedance
          B. To make the load impedance much less than the source
             impedance
          C. To use a balun at the antenna feed point
          D. To make the load impedance equal the source impedance

3E-6.3   What occurs when the impedance of an electrical load is equal to
          the internal impedance of the power source?
          A. The source delivers minimum power to the load
          B. There will be a high SWR condition
          C. No current can flow through the circuit
          D. The source delivers maximum power to the load

3E-6.4   Why is *impedance matching* important in radio work?
          A. So the source can deliver maximum power to the load
          B. So the load will draw minimum power from the source
          C. To ensure that there is less resistance than reactance in the
             circuit
          D. To ensure that the resistance and reactance in the circuit are
             equal

3E-7.2   What is the unit measurement of reactance?
          A. Mho
          B. Ohm
          C. Ampere
          D. Siemens

3E-7.4   What is the unit measurement of impedance?
          A. Ohm
          B. Volt
          C. Ampere
          D. Watt

3E-10.1   What is a *bel*?
          A. The basic unit used to describe a change in power levels
          B. The basic unit used to describe a change in inductances
          C. The basic unit used to describe a change in capacitances
          D. The basic unit used to describe a change in resistances

3E-10.2   What is a *decibel*?
          A. A unit used to describe a change in power levels, equal to
             0.1 bel
          B. A unit used to describe a change in power levels, equal to
             0.01 bel
          C. A unit used to describe a change in power levels, equal to
             10 bels
          D. A unit used to describe a change in power levels, equal to
             100 bels

3E-10.3   Under ideal conditions, a barely detectable change in loudness is
            approximately how many dB?
            A. 12 dB
            B. 6 dB
            C. 3 dB
            D. 1 dB

3E-10.4   A two-times increase in power results in a change of how many dB?
            A. Multiplying the original power by 2 gives a new power that is
               1 dB higher
            B. Multiplying the original power by 2 gives a new power that is
               3 dB higher
            C. Multiplying the original power by 2 gives a new power that is
               6 dB higher
            D. Multiplying the original power by 2 gives a new power that is
               12 dB higher

3E-10.5   An increase of 6 dB results from raising the power by how many
            times?
            A. Multiply the original power by 1.5 to get the new power
            B. Multiply the original power by 2 to get the new power
            C. Multiply the original power by 3 to get the new power
            D. Multiply the original power by 4 to get the new power

3E-10.6   A decrease of 3 dB results from lowering the power by how many
            times?
            A. Divide the original power by 1.5 to get the new power
            B. Divide the original power by 2 to get the new power
            C. Divide the original power by 3 to get the new power
            D. Divide the original power by 4 to get the new power

3E-10.7   A signal strength report is "10 dB over S9." If the transmitter power
            is reduced from 1500 watts to 150 watts, what should be the new
            signal strength report?
            A. S5
            B. S7
            C. S9
            D. S9 plus 5 dB

3E-10.8   A signal strength report is "20 dB over S9." If the transmitter power
            is reduced from 1500 watts to 150 watts, what should be the new
            signal strength report?
            A. S5
            B. S7
            C. S9
            D. S9 plus 10 dB

3E-10.9   A signal strength report is "20 dB over S9." If the transmitter power
            is reduced from 1500 watts to 15 watts, what should be the new
            signal strength report?
            A. S5
            B. S7
            C. S9
            D. S9 plus 10 dB

3E-12.1   If a 1.0-ampere current source is connected to two parallel-connected 10 ohm resistors, how much current passes through each resistor?
A. 10 amperes
B. 2 amperes
C. 1 ampere
D. 0.5 ampere

3E-12.3   In a parallel circuit with a voltage source and several branch resistors, what relationship does the total current have to the current in the branch circuits?
A. The total current equals the average of the branch current through each resistor
B. The total current equals the sum of the branch current through each resistor
C. The total current decreases as more parallel resistors are added to the circuit
D. The total current is calculated by adding the voltage drops across each resistor and multiplying the sum by the total number of all circuit resistors

3E-13.1   How many watts of electrical power are being used when a 400-vdc power source supplies an 800 ohm load?
A. 0.5 watt
B. 200 watts
C. 400 watts
D. 320,000 watts

3E-13.2   How many watts of electrical power are being consumed by a 12-vdc pilot light which draws 0.2-amperes?
A. 60 watts
B. 24 watts
C. 6 watts
D. 2.4 watts

3E-13.3   How many watts are being dissipated when 7.0-milliamperes flows through 1.25 kilohms?
A. Approximately 61 milliwatts
B. Approximately 39 milliwatts
C. Approximately 11 milliwatts
D. Approximately 9 milliwatts

3E-14.1   How is the total resistance calculated for several resistors in series?
A. The total resistance must be divided by the number of resistors to ensure accurate measurement of resistance
B. The total resistance is always the lowest-rated resistance
C. The total resistance is found by adding the individual resistances together
D. The tolerance of each resistor must be raised proportionally to the number of resistors

3E-14.2   What is the total resistance of two equal, parallel-connected resistors?
A. Twice the resistance of either resistance
B. The sum of the two resistances
C. The total resistance cannot be determined without knowing the exact resistances
D. Half the resistance of either resistor

3E-14.3   What is the total inductance of two equal, parallel-connected inductors?
           A. Half the inductance of either inductor, assuming no mutual coupling
           B. Twice the inductance of either inductor, assuming no mutual coupling
           C. The sum of the two inductances, assuming no mutual coupling
           D. The total inductance cannot be determined without knowing the exact inductances

3E-14.4   What is the total capacitance of two equal, parallel-connected capacitors?
           A. Half the capacitance of either capacitor
           B. Twice the capacitance of either capacitor
           C. The value of either capacitor
           D. The total capacitance cannot be determined without knowing the exact capacitances

3E-14.5   What is the total resistance of two equal, series-connected resistors?
           A. Half the resistance of either resistor
           B. Twice the resistance of either resistor
           C. The value of either resistor
           D. The total resistance cannot be determined without knowing the exact resistances

3E-14.6   What is the total inductance of two equal, series-connected inductors?
           A. Half the inductance of either inductor, assuming no mutual coupling
           B. Twice the inductance of either inductor, assuming no mutual coupling
           C. The value of either inductor, assuming no mutual coupling
           D. The total inductance cannot be determined without knowing the exact inductances

3E-14.7   What is the total capacitance of two equal, series-connected capacitors?
           A. Half the capacitance of either capacitor
           B. Twice the capacitance of either capacitor
           C. The value of either capacitor
           D. The total capacitance cannot be determined without knowing the exact capacitances

3E-15.1   What is the voltage across a 500 turn secondary winding in a transformer when the 2250 turn primary is connected to 117-vac?
           A. 2369 volts
           B. 526.5 volts
           C. 26 volts
           D. 5.8 volts

3E-15.2   What is the turns ratio of a transformer to match an audio amplifier having an output impedance of 200 ohms to a speaker having an impedance of 10 ohms?
           A. 4.47 to 1
           B. 14.14 to 1
           C. 20 to 1
           D. 400 to 1

3E-15.3   What is the turns ratio of a transformer to match an audio amplifier having an output impedance of 600 ohms to a speaker having an impedance of 4 ohms?
A. 12.2 to 1
B. 24.4 to 1
C. 150 to 1
D. 300 to 1

3E-15.4   What is the impedance of a speaker which requires a transformer with a turns ratio of 24 to 1 to match an audio amplifier having an output impedance of 2000 ohms?
A. 576 ohms
B. 83.3 ohms
C. 7.0 ohms
D. 3.5 ohms

3E-16.1   What is the voltage that would produce the same amount of heat over time in a resistive element as would an applied sine wave ac voltage?
A. A dc voltage equal to the peak-to-peak value of the ac voltage
B. A dc voltage equal to the RMS value of the ac voltage
C. A dc voltage equal to the average value of the ac voltage
D. A dc voltage equal to the peak value of the ac voltage

3E-16.2   What is the peak-to-peak voltage of a sine wave which has an RMS voltage of 117-volts?
A. 82.7 volts
B. 165.5 volts
C. 183.9 volts
D. 330.9 volts

3E-16.3   A sine wave of 17-volts peak is equivalent to how many volts RMS?
A. 8.5 volts
B. 12 volts
C. 24 volts
D. 34 volts

SUBELEMENT 3BF—Circuit Components (1 question)

3F-1.5   What is the effect of an increase in ambient temperature on the
         resistance of a carbon resistor?
         A. The resistance will increase by 20% for every 10 degrees
            centigrade that the temperature increases
         B. The resistance stays the same
         C. The resistance change depends on the resistor's temperature
            coefficient rating
         D. The resistance becomes time dependent

3F-2.6   What type of capacitor is often used in power supply circuits to filter
         the rectified ac?
         A. Disc ceramic
         B. Vacuum variable
         C. Mica
         D. Electrolytic

3F-2.7   What type of capacitor is used in power supply circuits to filter
         transient voltage spikes across the transformer secondary winding?
         A. High-value
         B. Trimmer
         C. Vacuum variable
         D. Suppressor

3F-3.5   How do inductors become self-resonant?
         A. Through distributed electromagnetism
         B. Through eddy currents
         C. Through distributed capacitance
         D. Through parasitic hysteresis

3F-4.1   What circuit component can change 120-vac to 400-vac?
         A. A transformer
         B. A capacitor
         C. A diode
         D. An SCR

3F-4.2   What is the source of energy connected to in a transformer?
         A. To the secondary winding
         B. To the primary winding
         C. To the core
         D. To the plates

3F-4.3   When there is no load attached to the secondary winding of a
         transformer, what is current in the primary winding called?
         A. Magnetizing current
         B. Direct current
         C. Excitation current
         D. Stabilizing current

3F-4.4   In what terms are the primary and secondary windings ratings of a
         power transformer usually specified?
         A. Joules per second
         B. Peak inverse voltage
         C. Coulombs per second
         D. Volts or volt-amperes

3F-5.1   What is the peak-inverse-voltage rating of a power supply rectifier?
- A. The highest transient voltage the diode will handle
- B. 1.4 times the ac frequency
- C. The maximum voltage to be applied in the non-conducting direction
- D. 2.8 times the ac frequency

3F-5.2   Why must silicon rectifier diodes be thermally protected?
- A. Because of their proximity to the power transformer
- B. Because they will be destroyed if they become too hot
- C. Because of their susceptibility to transient voltages
- D. Because of their use in high-voltage applications

3F-5.4   What are the two major ratings for silicon diode rectifiers of the type used in power supply circuits which must not be exceeded?
- A. Peak load impedance; peak voltage
- B. Average power; average voltage
- C. Capacitive reactance; avalanche voltage
- D. Peak inverse voltage; average forward current

SUBELEMENT 3BG—Practical Circuits (1 question)

3G-1.1   Why should a resistor and capacitor be wired in parallel with power supply rectifier diodes?
A. To equalize voltage drops and guard against transient voltage spikes
B. To ensure that the current through each diode is about the same
C. To smooth the output waveform
D. To decrease the output voltage

3G-1.2   What function do capacitors serve when resistors and capacitors are connected in parallel with high voltage power supply rectifier diodes?
A. They double or triple the output voltage
B. They block the alternating current
C. They protect those diodes that develop back resistance faster than other diodes
D. They regulate the output voltage

3G-1.3   What is the output waveform of an unfiltered full-wave rectifier connected to a resistive load?
A. A steady dc voltage
B. A sine wave at half the frequency of the ac input
C. A series of pulses at the same frequency as the ac input
D. A series of pulses at twice the frequency of the ac input

3G-1.4   How many degrees of each cycle does a half-wave rectifier utilize?
A. 90 degrees
B. 180 degrees
C. 270 degrees
D. 360 degrees

3G-1.5   How many degrees of each cycle does a full-wave rectifier utilize?
A. 90 degrees
B. 180 degrees
C. 270 degrees
D. 360 degrees

3G-1.6   Where is a power supply bleeder resistor connected?
A. Across the filter capacitor
B. Across the power-supply input
C. Between the transformer primary and secondary
D. Across the inductor in the output filter

3G-1.7   What components comprise a power supply filter network?
A. Diodes
B. Transformers and transistors
C. Quartz crystals
D. Capacitors and inductors

3G-1.8   What should be the peak-inverse-voltage rating of the rectifier in a full-wave power supply?
A. One-quarter the normal output voltage of the power supply
B. Half the normal output voltage of the power supply
C. Equal to the normal output voltage of the power supply
D. Double the normal peak output voltage of the power supply

3G-1.9   What should be the peak-inverse-voltage rating of the rectifier in a half-wave power supply?
A. One-quarter to one-half the normal peak output voltage of the power supply
B. Half the normal output voltage of the power supply
C. Equal to the normal output voltage of the power supply
D. One to two times the normal peak output voltage of the power supply

3G-2.8   What should the impedance of a low-pass filter be as compared to the impedance of the transmission line into which it is inserted?
A. Substantially higher
B. About the same
C. Substantially lower
D. Twice the transmission line impedance

SUBELEMENT 3BH—Signals and Emissions (2 questions)

3H-2.1   What is the term for alteration of the amplitude of an rf wave for the purpose of conveying information?
A. Frequency modulation
B. Phase modulation
C. Amplitude rectification
D. Amplitude modulation

3H-2.3   What is the term for alteration of the phase of an rf wave for the purpose of conveying information?
A. Pulse modulation
B. Phase modulation
C. Phase rectification
D. Amplitude modulation

3H-2.4   What is the term for alteration of the frequency of an rf wave for the purpose of conveying information?
A. Phase rectification
B. Frequency rectification
C. Amplitude modulation
D. Frequency modulation

3H-3.1   In what emission type does the instantaneous amplitude (envelope) of the rf signal vary in accordance with the modulating af?
A. Frequency shift keying
B. Pulse modulation
C. Frequency modulation
D. Amplitude modulation

3H-3.2   What determines the spectrum space occupied by each group of sidebands generated by a correctly operating emission A3E transmitter?
A. The audio frequencies used to modulate the transmitter
B. The phase angle between the audio and radio frequencies being mixed
C. The radio frequencies used in the transmitter's VFO
D. The CW keying speed

3H-4.1   How much is the carrier suppressed in an emission J3E transmission?
A. No more than 20 dB below peak output power
B. No more than 30 dB below peak output power
C. At least 40 dB below peak output power
D. At least 60 dB below peak output power

3H-4.2   What is one advantage of carrier suppression in an emission A3E transmission?
A. Only half the bandwidth is required for the same information content
B. Greater modulation percentage is obtainable with lower distortion
C. More power can be put into the sidebands
D. Simpler equipment can be used to receive a double-sideband suppressed-carrier signal

3H-5.1   Which one of the telephony emissions popular with amateurs occupies the narrowest band of frequencies?
A. Single-sideband emission
B. Double-sideband emission
C. Phase-modulated emission
D. Frequency-modulated emission

3H-5.2   Which emission type is produced by a telephony transmitter having a balanced modulator followed by a 2.5-kHz bandpass filter?
A. PM
B. AM
C. SSB
D. FM

3H-7.2   What emission is produced by a reactance modulator connected to an rf power amplifier?
A. Multiplex modulation
B. Phase modulation
C. Amplitude modulation
D. Pulse modulation

3H-8.1   What purpose does the carrier serve in an emission A3E transmission?
A. The carrier separates the sidebands so they don't cancel in the receiver
B. The carrier contains the modulation information
C. The carrier maintains symmetry of the sidebands to prevent distortion
D. The carrier serves as a reference signal for demodulation by an envelope detector

3H-8.2   What signal component appears in the center of the frequency band of an emission A3E transmission?
A. The lower sidebands
B. The subcarrier
C. The carrier
D. The pilot tone

3H-9.1   What sidebands are generated by an emission A3E transmitter with a 7250-kHz carrier modulated less than 100% by an 800-Hz pure sine wave?
A. 7250.8 kHz and 7251.6 kHz
B. 7250.0 kHz and 7250.8 kHz
C. 7249.2 kHz and 7250.8 kHz
D. 7248.4 kHz and 7249.2 kHz

3H-10.1   How many times over the maximum deviation is the bandwidth of an emission F3E transmission?
A. 1.5
B. At least 2.0
C. At least 4.0
D. The bandwidth cannot be determined without knowing the exact carrier and modulating frequencies involved

3H-10.2   What is the total bandwidth of an emission F3E transmission having 5-kHz deviation and 3-kHz af?
A. 3 kHz
B. 5 kHz
C. 8 kHz
D. 16 kHz

3H-11.1 What happens to the shape of the rf envelope, as viewed on an oscilloscope, of an emission A3E transmission?
A. The amplitude of the envelope increases and decreases in proportion to the modulating signal
B. The amplitude of the envelope remains constant
C. The brightness of the envelope increases and decreases in proportion to the modulating signal
D. The frequency of the envelope increases and decreases in proportion to the amplitude of the modulating signal

3H-13.1 What results when an emission J3E transmitter is overmodulated?
A. The signal becomes louder with no other effects
B. The signal occupies less bandwidth with poor high frequency response
C. The signal has higher fidelity and improved signal-to-noise ratio
D. The signal becomes distorted and occupies more bandwidth

3H-13.2 What results when an emission A3E transmitter is overmodulated?
A. The signal becomes louder with no other effects
B. The signal becomes distorted and occupies more bandwidth
C. The signal occupies less bandwidth with poor high frequency response
D. The transmitter's carrier frequency deviates

3H-15.1 What is the frequency deviation for a 12.21-MHz reactance-modulated oscillator in a 5-kHz deviation, 146.52-MHz F3E transmitter?
A. 41.67 Hz
B. 416.7 Hz
C. 5 kHz
D. 12 kHz

3H-15.2 What stage in a transmitter would translate a 5.3-MHz input signal to 14.3-MHz?
A. A mixer
B. A beat frequency oscillator
C. A frequency multiplier
D. A linear translator stage

3H-16.4 How many frequency components are in the signal from an af shift keyer at any instant?
A. One
B. Two
C. Three
D. Four

3H-16.5 How is frequency shift related to keying speed in an fsk signal?
A. The frequency shift in hertz must be at least four times the keying speed in WPM
B. The frequency shift must not exceed 15 Hz per WPM of keying speed
C. Greater keying speeds require greater frequency shifts
D. Greater keying speeds require smaller frequency shifts

3I-1.3 Why is a Yagi antenna often used for radiocommunications on the 20 meter band?
  A. It provides excellent omnidirectional coverage in the horizontal plane
  B. It is smaller, less expensive and easier to erect than a dipole or vertical antenna
  C. It discriminates against interference from other stations off to the side or behind
  D. It provides the highest possible angle of radiation for the HF bands

3I-1.7 What method is best suited to match an unbalanced coaxial feed line to a Yagi antenna?
  A. "T" match
  B. Delta match
  C. Hairpin match
  D. Gamma match

3I-1.9 How can the bandwidth of a parasitic beam antenna be increased?
  A. Use larger diameter elements
  B. Use closer element spacing
  C. Use traps on the elements
  D. Use tapered-diameter elements

3I-2.1 How much gain over a half-wave dipole can a two-element cubical quad antenna provide?
  A. Approximately 0.6 dB
  B. Approximately 2 dB
  C. Approximately 6 dB
  D. Approximately 12 dB

3I-3.1 How long is each side of a cubical quad antenna driven element for 21.4-MHz?
  A. 1.17 feet
  B. 11.7 feet
  C. 47 feet
  D. 469 feet

3I-3.2 How long is each side of a cubical quad antenna driven element for 14.3-MHz?
  A. 1.75 feet
  B. 17.6 feet
  C. 23.4 feet
  D. 70.3 feet

3I-3.3 How long is each side of a cubical quad antenna reflector element for 29.6-MHz?
  A. 8.23 feet
  B. 8.7 feet
  C. 9.7 feet
  D. 34.8 feet

3I-3.4 How long is each leg of a symmetrical delta loop antenna driven element for 28.7-MHz?
  A. 8.75 feet
  B. 11.32 feet
  C. 11.7 feet
  D. 35 feet

3I-3.5   How long is each leg of a symmetrical delta loop antenna driven element for 24.9-MHz?
- A. 10.09 feet
- B. 13.05 feet
- C. 13.45 feet
- D. 40.36 feet

3I-3.6   How long is each leg of a symmetrical delta loop antenna reflector element for 14.1-MHz?
- A. 18.26 feet
- B. 23.76 feet
- C. 24.35 feet
- D. 73.05 feet

3I-3.7   How long is the driven element of a Yagi antenna for 14.0-MHz?
- A. Approximately 17 feet
- B. Approximately 33 feet
- C. Approximately 35 feet
- D. Approximately 66 feet

3I-3.8   How long is the director element of a Yagi antenna for 21.1-MHz?
- A. Approximately 42 feet
- B. Approximately 21 feet
- C. Approximately 17 feet
- D. Approximately 10.5 feet

3I-3.9   How long is the reflector element of a Yagi antenna for 28.1-MHz?
- A. Approximately 8.75 feet
- B. Approximately 16.6 feet
- C. Approximately 17.5 feet
- D. Approximately 35 feet

3I-5.1   What is the feed-point impedance for a half-wavelength dipole HF antenna suspended horizontally one-quarter wavelength or more above the ground?
- A. Approximately 50 ohms, resistive
- B. Approximately 73 ohms, resistive and inductive
- C. Approximately 50 ohms, resistive and capacitive
- D. Approximately 73 ohms, resistive

3I-5.2   What is the feed-point impedance of a quarter-wavelength vertical HF antenna with a horizontal ground plane?
- A. Approximately 18 ohms
- B. Approximately 36 ohms
- C. Approximately 52 ohms
- D. Approximately 72 ohms

3I-5.3   What is an advantage of downward sloping radials on a ground-plane antenna?
- A. Sloping the radials downward lowers the radiation angle
- B. Sloping the radials downward brings the feed-point impedance close to 300 ohms
- C. Sloping the radials downward allows rainwater to run off the antenna
- D. Sloping the radials downward brings the feed-point impedance closer to 50 ohms

3I-5.4   What happens to the feed-point impedance of a ground-plane antenna when the radials slope downward from the base of the antenna?
A. The feed-point impedance decreases
B. The feed-point impedance increases
C. The feed-point impedance stays the same
D. The feed-point impedance becomes purely capacitive

3I-6.1   Compared to a dipole antenna, what are the directional radiation characteristics of a cubical quad HF antenna?
A. The quad has more directivity in the horizontal plane but less directivity in the vertical plane
B. The quad has less directivity in the horizontal plane but more directivity in the vertical plane
C. The quad has more directivity in both horizontal and vertical planes
D. The quad has less directivity in both horizontal and vertical planes

3I-6.2   What is the radiation pattern of an ideal half-wavelength dipole HF antenna?
A. If it is installed parallel to the earth, it radiates well in a figure-eight pattern at right angles to the antenna wire
B. If it is installed parallel to the earth, it radiates well in a figure-eight pattern off both ends of the antenna wire
C. If it is installed parallel to the earth, it radiates equally well in all directions
D. If it is installed parallel to the earth, the pattern will have two lobes on one side of the antenna wire, and one larger lobe on the other side

3I-6.3   How does proximity to the ground affect the radiation pattern of a horizontal dipole HF antenna?
A. If the antenna is too far from the ground, the pattern becomes unpredictable
B. If the antenna is less than one-half wavelength from the ground, reflected radio waves from the ground distort the radiation pattern of the antenna
C. A dipole antenna's radiation pattern is unaffected by its distance to the ground
D. If the antenna is less than one-half wavelength from the ground, radiation off the ends of the wire is reduced

3I-6.4   What does the term *antenna front-to-back ratio* mean?
A. The number of directors versus the number of reflectors
B. The relative position of the driven element with respect to the reflectors and directors
C. The power radiated in the major radiation lobe compared to the power radiated in exactly the opposite direction
D. The power radiated in the major radiation lobe compared to the power radiated 90 degrees away from that direction

3I-6.5 What effect upon the radiation pattern of an HF dipole antenna will a slightly smaller parasitic parallel element located a few feet away in the same horizontal plane have?
  A. The radiation pattern will not change appreciably
  B. A major lobe will develop in the horizontal plane, parallel to the two elements
  C. A major lobe will develop in the vertical plane, away from the ground
  D. If the spacing is greater than 0.1 wavelength, a major lobe will develop in the horizontal plane to the side of the driven element toward the parasitic element

3I-6.6 What is the meaning of the term *main lobe* as used in reference to a directional antenna?
  A. The direction of least radiation from an antenna
  B. The point of maximum current in a radiating antenna element
  C. The direction of maximum radiated field strength from a radiating antenna
  D. The maximum voltage standing wave point on a radiating element

3I-7.1 Upon what does the characteristic impedance of a parallel-conductor antenna feed line depend?
  A. The distance between the centers of the conductors and the radius of the conductors
  B. The distance between the centers of the conductors and the length of the line
  C. The radius of the conductors and the frequency of the signal
  D. The frequency of the signal and the length of the line

3I-7.2 What is the characteristic impedance of various coaxial cables commonly used for antenna feed lines at amateur stations?
  A. Around 25 and 30 ohms
  B. Around 50 and 75 ohms
  C. Around 80 and 100 ohms
  D. Around 500 and 750 ohms

3I-7.3 What effect, if any, does the length of a coaxial cable have upon its characteristic impedance?
  A. The length has no effect on the characteristic impedance
  B. The length affects the characteristic impedance primarily above 144 MHz
  C. The length affects the characteristic impedance primarily below 144 MHz
  D. The length affects the characteristic impedance at any frequency

3I-7.4 What is the characteristic impedance of flat-ribbon TV-type twinlead?
  A. 50 ohms
  B. 75 ohms
  C. 100 ohms
  D. 300 ohms

3I-8.4   What is the cause of power being reflected back down an antenna feed line?
- A. Operating an antenna at its resonant frequency
- B. Using more transmitter power than the antenna can handle
- C. A difference between feed line impedance and antenna feed-point impedance
- D. Feeding the antenna with unbalanced feed line

3I-9.3   What will be the standing wave ratio when a 50 ohm feed line is connected to a resonant antenna having a 200 ohm feed-point impedance?
- A. 4:1
- B. 1:4
- C. 2:1
- D. 1:2

3I-9.4   What will be the standing wave ratio when a 50 ohm feed line is connected to a resonant antenna having a 10 ohm feed-point impedance?
- A. 2:1
- B. 50:1
- C. 1:5
- D. 5:1

3I-9.5   What will be the standing wave ratio when a 50 ohm feed line is connected to a resonant antenna having a 50 ohm feed-point impedance?
- A. 2:1
- B. 50:50
- C. 1:1
- D. 0:0

3I-11.1   How does the characteristic impedance of a coaxial cable affect the amount of attenuation to the rf signal passing through it?
- A. The attenuation is affected more by the characteristic impedance at frequencies above 144 MHz than at frequencies below 144 MHz
- B. The attenuation is affected less by the characteristic impedance at frequencies above 144 MHz than at frequencies below 144 MHz
- C. The attenuation related to the characteristic impedance is about the same at all amateur frequencies below 1.5 GHz
- D. The difference in attenuation depends on the emission type in use

3I-11.2   How does the amount of attenuation to a 2 meter signal passing through a coaxial cable differ from that to a 160 meter signal?
- A. The attenuation is greater at 2 meters
- B. The attenuation is less at 2 meters
- C. The attenuation is the same at both frequencies
- D. The difference in attenuation depends on the emission type in use

3I-11.4   What is the effect on its attenuation when flat-ribbon TV-type twinlead is wet?
- A. Attenuation decreases slightly
- B. Attenuation remains the same
- C. Attenuation decreases sharply
- D. Attenuation increases

3I-11.7   Why might silicone grease or automotive car wax be applied to
           flat-ribbon TV-type twinlead?
           A. To reduce "skin effect" losses on the conductors
           B. To reduce the buildup of dirt and moisture on the feed line
           C. To increase the velocity factor of the feed line
           D. To help dissipate heat during high-SWR operation

3I-11.8   In what values are rf feed line losses usually expressed?
           A. Bels/1000 ft
           B. dB/1000 ft
           C. Bels/100 ft
           D. dB/100 ft

3I-11.10   As the operating frequency increases, what happens to the
           dielectric losses in a feed line?
           A. The losses decrease
           B. The losses decrease to zero
           C. The losses remain the same
           D. The losses increase

3I-11.12   As the operating frequency decreases, what happens to the
           dielectric losses in a feed line?
           A. The losses decrease
           B. The losses increase
           C. The losses remain the same
           D. The losses become infinite

3I-12.1   What condition must be satisfied to prevent standing waves of
           voltage and current on an antenna feed line?
           A. The antenna feed point must be at dc ground potential
           B. The feed line must be an odd number of electrical quarter
              wavelengths long
           C. The feed line must be an even number of physical half
              wavelengths long
           D. The antenna feed-point impedance must be matched to the
              characteristic impedance of the feed line

3I-12.2   How is an inductively-coupled matching network used in an antenna
           system consisting of a center-fed resonant dipole and coaxial feed
           line?
           A. An inductively coupled matching network is not normally used
              in a resonant antenna system
           B. An inductively coupled matching network is used to increase
              the SWR to an acceptable level
           C. An inductively coupled matching network can be used to
              match the unbalanced condition at the transmitter output to
              the balanced condition required by the coaxial line
           D. An inductively coupled matching network can be used at the
              antenna feed point to tune out the radiation resistance

3I-12.5   What is an antenna-transmission line *mismatch*?
           A. A condition where the feed-point impedance of the antenna
              does not equal the output impedance of the transmitter
           B. A condition where the output impedance of the transmitter
              does not equal the characteristic impedance of the feed line
           C. A condition where a half-wavelength antenna is being fed with
              a transmission line of some length other than one-quarter
              wavelength at the operating frequency
           D. A condition where the characteristic impedance of the feed
              line does not equal the feed-point impedance of the antenna

Element 3B Answer Key

SUBELEMENT 3BA

3A-3.2	A
3A-3.3	A
3A-3.4	C
3A-3.5	C
3A-3.7	A
3A-4.1	C
3A-4.3	C
3A-6.1	B
3A-6.2	C
3A-6.6	A
3A-8.6	D
3A-9.1	C
3A-9.2	A
3A-9.3	D
3A-9.4	A
3A-9.5	B
3A-9.6	C
3A-9.7	A
3A-9.8	A
3A-9.9	C
3A-9.10	B
3A-9.11	C
3A-9.12	A
3A-9.13	B
3A-9.14	C
3A-9.15	C
3A-9.16	C
3A-10.1	A
3A-10.2	C
3A-10.3	D
3A-10.4	C
3A-10.5	B
3A-10.6	C
3A-10.7	C
3A-10.8	C
3A-13.1	C
3A-13.2	D
3A-14.3	B
3A-14.4	**
3A-14.6	A
3A-15.1	D
3A-15.3	C
3A-15.4	B
3A-16.1	C
3A-16.2	B
3A-16.3	A
3A-16.4	A

SUBELEMENT 3BB

3B-1.4	C	p.2-1
3B-1.5	B	p.2-1
3B-2.1	B	p.2-7
3B-2.2	A	p.2-7
3B-2.3	C	p.2-7
3B-2.4	A	p.2-8
3B-2.6	B	p.2-7
3B-2.10	C	p.2-8
3B-2.11	D	p.2-8
3B-2.12	B	p.2-8
3B-3.8	A	p.2-6
3B-3.12	A	p.2-6
3B-4.1	A	p.2-2
3B-4.2	B	p.2-2
3B-5.1	D	p.2-11
3B-5.2	C	p.2-11
3B-6.1	B	p.2-12
3B-6.2	B	p.2-12
3B-6.3	B	p.2-12
3B-7.1	B	p.2-15
3B-7.2	A	p.2-15
3B-7.3	A	p.2-15
3B-7.4	C	p.2-13
3B-7.5	C	p.2-13
3B-8.1	C	p.2-13
3B-8.2	B	p.2-13
3B-8.3	B	p.2-13
3B-8.4	C	p.2-13
3B-8.5	C	p.2-15
3B-8.6	B	p.2-13
3B-8.7	C	p.2-13
3B-8.8	C	p.2-13
3B-8.9	C	p.2-13
3B-10.1	A	p.2-17
3B-10.2	B	p.2-17

SUBELEMENT 3BC

3C-1.6	C	p.3-3
3C-1.7	B	p.3-3
3C-1.9	B	p.3-3
3C-1.10	A	p.3-3

**This number was skipped when the FCC made up the question pool. There is no question number 3A-14.4.

3C-1.13	D	p.3-4		3D-9.2	A	p.4-11
3C-2.3	C	p.3-3		3D-9.3	D	p.4-11
3C-2.4	C	p.3-3		3D-10.1	B	p.4-19
3C-3.3	B	p.3-5		3D-10.2	B	p.4-19
3C-3.4	C	p.3-5		3D-10.3	C	p.4-19
3C-5.1	B	p.3-6		3D-10.4	C	p.4-19
3C-5.2	A	p.3-6		3D-10.5	A	p.4-19
3C-5.3	B	p.3-6		3D-12.2	D	p.4-16
3C-5.4	C	p.3-6		3D-12.3	B	p.4-16
3C-5.5	A	p.3-6		3D-12.4	C	p.4-16
3C-6.2	B	p.3-10		3D-12.5	D	p.4-16
3C-6.4	D	p.3-10		3D-13.1	A	p.4-15
3C-6.5	B	p.3-9		3D-13.2	C	p.4-15
3C-6.6	D	p.3-9		3D-13.3	D	p.4-15
3C-7.1	B	p.3-5		3D-14.6	B	p.4-18
3C-7.2	D	p.3-6		3D-14.7	C	p.4-18
3C-7.3	A	p.3-6		3D-15.1	B	p.4-12
3C-7.4	D	p.3-6		3D-15.2	A	p.4-12
3C-7.5	D	p.3-6		3D-15.3	B	p.4-8
3C-7.6	A	p.3-6		3D-15.4	D	p.4-8
3C-7.7	D	p.3-6		3D-17.2	A	p.4-15
3C-7.8	C	p.3-6		3D-17.3	C	p.4-15
3C-10.1	D	p.3-7		3D-17.4	A	p.4-15
3C-10.2	A	p.3-7		3D-17.5	B	p.4-15
3C-10.3	B	p.3-7		3D-17.6	C	p.4-15
3C-10.4	D	p.3-7				

SUBELEMENT 3BE

3E-1.1	C	p.5-23
3E-1.2	C	p.5-23
3E-3.1	B	p.5-21
3E-3.2	D	p.5-21
3E-3.3	D	p.5-21
3E-3.4	D	p.5-22
3E-3.5	A	p.5-21
3E-6.1	A	p.5-24
3E-6.2	D	p.5-24
3E-6.3	D	p.5-24
3E-6.4	A	p.5-24
3E-7.2	B	p.5-21
3E-7.4	A	p.5-23
3E-10.1	A	p.5-11
3E-10.2	A	p.5-11
3E-10.3	D	p.5-11
3E-10.4	B	p.5-11
3E-10.5	D	p.5-11
3E-10.6	B	p.5-11
3E-10.7	C	p.5-11
3E-10.8	D	p.5-11
3E-10.9	C	p.5-11
3E-12.1	D	p.5-5
3E-12.3	B	p.5-5
3E-13.1	B	p.5-9
3E-13.2	D	p.5-9

SUBELEMENT 3BD

3D-1.5	A	p.4-1
3D-1.6	A	p.4-4
3D-1.7	D	p.4-4
3D-1.8	C	p.4-2
3D-1.9	D	p.4-4
3D-2.4	B	p.4-6
3D-3.1	C	p.4-11
3D-3.2	C	p.4-11
3D-3.3	C	p.4-11
3D-3.4	D	p.4-11
3D-3.5	D	p.4-11
3D-4.1	A	p.4-13
3D-4.2	D	p.4-13
3D-4.3	C	p.4-13
3D-4.4	B	p.4-13
3D-4.5	B	p.4-13
3D-5.1	B	p.8-7
3D-5.5	A	p.8-7
3D-5.6	B	p.8-7
3D-5.7	B	p.8-7
3D-6.1	D	p.4-6
3D-6.2	A	p.4-6
3D-6.3	D	p.4-6
3D-6.4	B	p.4-6
3D-9.1	C	p.4-11

3E-13.3	A	p.5-9
3E-14.1	C	p.5-5
3E-14.2	D	p.5-5
3E-14.3	A	p.5-18
3E-14.4	B	p.5-14
3E-14.5	B	p.5-5
3E-14.6	B	p.5-18
3E-14.7	A	p.5-14
3E-15.1	C	p.5-19
3E-15.2	A	p.5-24
3E-15.3	A	p.5-24
3E-15.4	D	p.5-24
3E-16.1	B	p.5-2
3E-16.2	D	p.5-2
3E-16.3	B	p.5-2

SUBELEMENT 3BF

3F-1.5	C	p.6-4
3F-2.6	D	p.6-7
3F-2.7	D	p.6-7
3F-3.5	C	p.6-11
3F-4.1	A	p.6-12
3F-4.2	B	p.6-12
3F-4.3	A	p.6-12
3F-4.4	D	p.6-12
3F-5.1	C	p.6-15
3F-5.2	B	p.6-13
3F-5.4	D	p.6-15

SUBELEMENT 3BG

3G-1.1	A	p.7-4
3G-1.2	C	p.7-4
3G-1.3	D	p.7-1
3G-1.4	B	p.7-2
3G-1.5	D	p.7-2
3G-1.6	A	p.7-6
3G-1.7	D	p.7-5
3G-1.8	D	p.7-2
3G-1.9	D	p.7-2
3G-2.8	B	p.7-7

SUBELEMENT 3BH

3H-2.1	D	p.8-2
3H-2.3	B	p.8-8
3H-2.4	D	p.8-8
3H-3.1	D	p.8-2
3H-3.2	A	p.8-3
3H-4.1	C	p.8-6
3H-4.2	C	p.8-6
3H-5.1	A	p.8-6
3H-5.2	C	p.8-6
3H-7.2	B	p.8-10
3H-8.1	D	p.8-3

3H-8.2	C	p.8-3
3H-9.1	C	p.8-3
3H-10.1	B	p.8-11
3H-10.2	D	p.8-11
3H-11.1	A	p.8-5
3H-13.1	D	p.8-4
3H-13.2	B	p.8-4
3H-15.1	B	p.8-12
3H-15.2	A	p.7-14
3H-16.4	A	p.8-14
3H-16.5	C	p.8-14

SUBELEMENT 3BI

3I-1.3	C	p.9-7
3I-1.7	D	p.9-7
3I-1.9	A	p.9-11
3I-2.1	C	p.9-9
3I-3.1	B	p.9-9
3I-3.2	B	p.9-9
3I-3.3	B	p.9-9
3I-3.4	C	p.9-9
3I-3.5	C	p.9-9
3I-3.6	C	p.9-9
3I-3.7	B	p.9-9
3I-3.8	B	p.9-9
3I-3.9	C	p.9-9
3I-5.1	D	p.9-3
3I-5.2	B	p.9-11
3I-5.3	D	p.9-12
3I-5.4	B	p.9-12
3I-6.1	C	p.9-9
3I-6.2	A	p.9-3
3I-6.3	B	p.9-3
3I-6.4	C	p.9-10
3I-6.5	D	p.9-5
3I-6.6	C	p.9-5
3I-7.1	A	p.9-13
3I-7.2	B	p.9-14
3I-7.3	A	p.9-14
3I-7.4	D	p.9-14
3I-8.4	C	p.9-16
3I-9.3	A	p.9-17
3I-9.4	D	p.9-17
3I-9.5	C	p.9-17
3I-11.1	C	p.9-14
3I-11.2	A	p.9-14
3I-11.4	D	p.9-14
3I-11.7	B	p.9-14
3I-11.8	D	p.9-13
3I-11.10	D	p.9-13
3I-11.12	A	p.9-13
3I-12.1	D	p.9-16
3I-12.2	A	p.9-18
3I-12.5	D	p.9-18

Appendix A

US Customary—Metric Conversion Factors

International System of Units (SI) — Metric Units

Prefix	Symbol		Multiplication Factor
exa	E	10^{18} =	1,000,000,000,000,000,000
peta	P	10^{15} =	1,000,000,000,000,000
tera	T	10^{12} =	1,000,000,000,000
giga	G	10^{9} =	1,000,000,000
mega	M	10^{6} =	1,000,000
kilo	k	10^{3} =	1,000
hecto	h	10^{2} =	100
deca	da	10^{1} =	10
(unit)		10^{0} =	1
deci	d	10^{-1} =	0.1
centi	c	10^{-2} =	0.01
milli	m	10^{-3} =	0.001
micro	μ	10^{-6} =	0.000001
nano	n	10^{-9} =	0.000000001
pico	p	10^{-12} =	0.000000000001
femto	f	10^{-15} =	0.000000000000001
atto	a	10^{-18} =	0.000000000000000001

Linear
1 meter (m) = 100 centimeters (cm) = 1000 millimeters (mm)

Area
$1 \text{ m}^2 = 1 \times 10^4 \text{ cm}^2 = 1 \times 10^6 \text{ mm}^2$

Volume
$1 \text{ m}^3 = 1 \times 10^6 \text{ cm}^3 = 1 \times 10^9 \text{ mm}^3$
$1 \text{ liter (l)} = 1000 \text{ cm}^3 = 1 \times 10^6 \text{ mm}^3$

Mass
1 kilogram (kg) = 1000 grams (g)
(Approximately the mass of 1 liter of water)
1 metric ton (or tonne) = 1000 kg

US Customary Units

Linear Units

12 inches (in) = 1 foot (ft)
36 inches = 3 feet = 1 yard (yd)
1 rod = 5½ yards = 16½ feet
1 statute mile = 1760 yards = 5280 feet
1 nautical mile = 6076.11549 feet

Area
$1 \text{ ft}^2 = 144 \text{ in}^2$
$1 \text{ yd}^2 = 9 \text{ ft}^2 = 1296 \text{ in}^2$
$1 \text{ rod}^2 = 30¼ \text{ yd}^2$
$1 \text{ acre} = 4840 \text{ yd}^2 = 43,560 \text{ ft}^2$
$1 \text{ acre} = 160 \text{ rod}^2$
$1 \text{ mile}^2 = 640 \text{ acres}$

Volume
$1 \text{ ft}^3 = 1728 \text{ in}^3$
$1 \text{ yd}^3 = 27 \text{ ft}^3$

Liquid Volume Measure
$1 \text{ fluid ounce (fl oz)} = 8 \text{ fluidrams} = 1.804 \text{ in}^3$
1 pint (pt) = 16 fl oz
$1 \text{ quart (qt)} = 2 \text{ pt} = 32 \text{ fl oz} = 57¾ \text{ in}^3$
1 gallon (gal) = 4 qt = 231 in³
1 barrel = 31½ gal

Dry Volume Measure
$1 \text{ quart (qt)} = 2 \text{ pints (pt)} = 67.2 \text{ in}^3$
1 peck = 8 qt

$1 \text{ bushel} = 4 \text{ pecks} = 2150.42 \text{ in}^3$

Avoirdupois Weight
1 dram (dr) = 27.343 grains (gr) or (gr a)
1 ounce (oz) = 437.5 gr
1 pound (lb) = 16 oz = 7000 gr
1 short ton = 2000 lb, 1 long ton = 2240 lb

Troy Weight
1 grain troy (gr t) = 1 grain avoirdupois
1 pennyweight (dwt) or (pwt) = 24 gr t
1 ounce troy (oz t) = 480 grains
1 lb t = 12 oz t = 5760 grains

Apothecaries' Weight
1 grain apothecaries' (gr ap) = 1 gr t = 1 gr a
1 dram ap (dr ap) = 60 gr
1 oz ap = 1 oz t = 8 dr ap = 480 gr
1 lb ap = 1 lb t = 12 oz ap = 5760 gr

Temperature
$°F = 9/5 °C + 32$
$°C = 5/9 (°F - 32)$
$K = °C + 273$
$°C = K - 273$

Multiply ⟶

Metric Unit = Conversion Factor × U S Customary Unit

⟵ **Divide**

Metric Unit ÷ Conversion Factor = U S Customary Unit

Metric Unit =	Conversion Factor ×	U S Unit	Metric Unit =	Conversion Factor ×	U S Unit
(Length)			(Volume)		
mm	25.4	inch	mm^3	16387.064	in^3
cm	2.54	inch	cm^3	16.387	in^3
cm	30.48	foot	m^3	0.028316	ft^3
m	0.3048	foot	m^3	0.764555	yd^3
m	0.9144	yard	ml	16.387	in^3
km	1.609	mile	ml	29.57	fl oz
km	1.852	nautical mile	ml	473	pint
(Area)			ml	946.333	quart
mm^2	645.16	$inch^2$	l	28.32	ft^3
cm^2	6.4516	in^2	l	0.9463	quart
cm^2	929.03	ft^2	l	3.785	gallon
m^2	0.0929	ft^2	l	1.101	dry quart
cm^2	8361.3	yd^2	l	8.809	peck
m^2	0.83613	yd^2	l	35.238	bushel
m^2	4047	acre			
km^2	2.59	mi^2			
(Mass)	(Avoirdupois Weight)		(Mass)	(Troy Weight)	
grams	0.0648	grains	g	31.103	oz t
g	28.349	oz	g	373.248	lb t
g	453.59	lb	(Mass)	(Apothecaries' Weight)	
kg	0.45359	lb	g	3.387	dr ap
tonne	0.907	short ton	g	31.103	oz ap
tonne	1.016	long ton	g	373.248	lb ap

Standard Resistance Values

Numbers in **bold** type are ± 10% values. Others are 5% values.

Ohms										Megohms				
1.0	3.6	**12**	43	**150**	510	**1800**	6200	**22000**	75000	0.24	0.62	1.6	4.3	11.0
1.1	**3.9**	13	**47**	160	**560**	2000	**6800**	24000	**82000**	**0.27**	**0.68**	**1.8**	**4.7**	**12.0**
1.2	4.3	**15**	51	**180**	620	**2200**	7500	**27000**	91000	0.30	0.75	2.0	5.1	13.0
1.3	**4.7**	16	**56**	200	**680**	2400	**8200**	30000	**100000**	**0.33**	**0.82**	**2.2**	**5.6**	**15.0**
1.5	5.1	**18**	62	**220**	750	**2700**	9100	**33000**	110000	0.36	0.91	2.4	6.2	16.0
1.6	**5.6**	20	**68**	240	**820**	3000	**10000**	36000	**120000**	**0.39**	**1.0**	**2.7**	**6.8**	**18.0**
1.8	6.2	**22**	75	**270**	910	**3300**	11000	**39000**	130000	0.43	1.1	3.0	7.5	20.0
2.0	**6.8**	24	**82**	300	**1000**	3600	**12000**	43000	**150000**	**0.47**	**1.2**	**3.3**	**8.2**	**22.0**
2.2	7.5	**27**	91	**330**	1100	**3900**	13000	**47000**	160000	0.51	1.3	3.6	9.1	
2.4	**8.2**	30	**100**	360	**1200**	4300	**15000**	51000	**180000**	**0.56**	**1.5**	**3.9**	**10.0**	
2.7	9.1	**33**	110	**390**	1300	**4700**	16000	**56000**	200000					
3.0	**10.0**	36	**120**	430	**1500**	5100	**18000**	62000	**220000**					
3.3	11.0	**39**	130	**470**	1600	**5600**	20000	**68000**						

Resistor Color Code

Color	Sig. Figure	Decimal Multiplier	Tolerance (%)	Color	Sig. Figure	Decimal Multiplier	Tolerance (%)
Black	0	1		Violet	7	10,000,000	
Brown	1	10		Gray	8	100,000,000	
Red	2	100		White	9	1.000,000,000	
Orange	3	1,000		Gold	—	0.1	5
Yellow	4	10,000		Silver	—	0.01	10
Green	5	100,000		No color	—		20
Blue	6	1,000,000					

Schematic Symbols Used in Circuit Diagrams

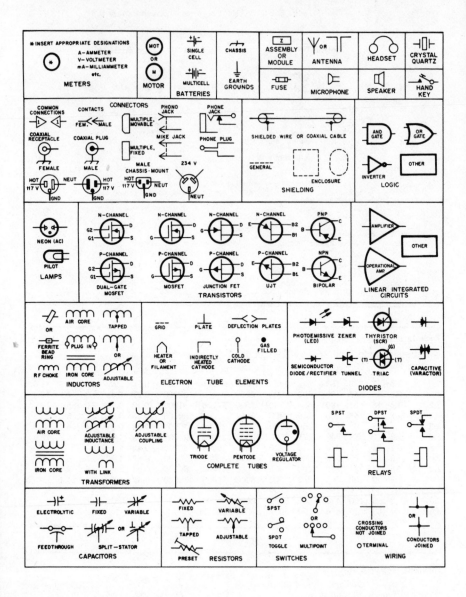

Standard Capacitance Values

pF	pF
0.3	470
5	500
6	510
6.8	560
7.5	600
8	680
10	750
12	800
15	820
18	910
20	1000
22	1000
24	1200
25	1200
27	1300
30	1500
33	1500
39	1600
47	1800
50	2000
51	2200
56	2500
68	2700
75	3000
82	3300
91	3900
100	4000
120	4300
130	4700
150	4700
180	5000
200	5000
220	5600
240	6800
250	7500
270	8200
300	10000
330	10000
350	20000
360	30000
390	40000
400	50000
470	

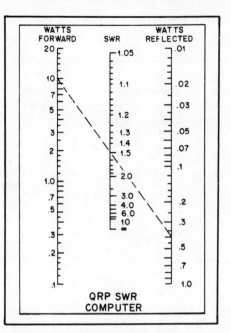

Nomograph of SWR versus forward and
reflected power for levels up to 20 watts.
Dashed line shows an SWR of 1.5:1 for 10 W
forward and 0.4 W reflected.

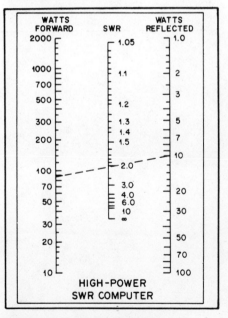

Nomograph of SWR versus forward and
reflected power for levels up to 2000 watts.
Dashed line shows an SWR of 2:1 for 90 W
forward and 10 W reflected.

Fractions of an Inch with Metric Equivalents

Fractions Of An Inch		Decimals Of An Inch	Millimeters	Fractions Of An Inch		Decimals Of An Inch	Millimeters
	1/64	0.0156	0.397		33/64	0.5156	13.097
1/32		0.0313	0.794	17/32		0.5313	13.494
	3/64	0.0469	1.191		35/64	0.5469	13.891
1/16		0.0625	1.588	9/16		0.5625	14.288
	5/64	0.0781	1.984		37/64	0.5781	14.684
3/32		0.0938	2.381	19/32		0.5938	15.081
	7/64	0.1094	2.778		39/64	0.6094	15.478
1/8		0.1250	3.175	5/8		0.6250	15.875
	9/64	0.1406	3.572		41/64	0.6406	16.272
5/32		0.1563	3.969	21/32		0.6563	16.669
	11/64	0.1719	4.366		43/64	0.6719	17.066
3/16		0.1875	4.763	11/16		0.6875	17.463
	13/64	0.2031	5.159		45/64	0.7031	17.859
7/32		0.2188	5.556	23/32		0.7188	18.256
	15/64	0.2344	5.953		47/64	0.7344	18.653
1/4		0.2500	6.350	3/4		0.7500	19.050
	17/64	0.2656	6.747		49/64	0.7656	19.447
9/32		0.2813	7.144	25/32		0.7813	19.844
	19/64	0.2969	7.541		51/64	0.7969	20.241
5/16		0.3125	7.938	13/16		0.8125	20.638
	21/64	0.3281	8.334		53/64	0.8281	21.034
11/32		0.3438	8.731	27/32		0.8438	21.431
	23/64	0.3594	9.128		55/64	0.8594	21.828
3/8		0.3750	9.525	7/8		0.8750	22.225
	25/64	0.3906	9.922		57/64	0.8906	22.622
13/32		0.4063	10.319	29/32		0.9063	23.019
	27/64	0.4219	10.716		59/64	0.9219	23.416
7/16		0.4375	11.113	15/16		0.9375	23.813
	29/64	0.4531	11.509		61/64	0.9531	24.209
15/32		0.4688	11.906	31/32		0.9688	24.606
	31/64	0.4844	12.303		63/64	0.9844	25.003
1/2		0.50000	12.700	1		1.0000	25.400

Appendix B

Equations Used in this Book

$$V_{peak} = V_{RMS} \times \sqrt{2} = V_{RMS} \times 1.414 \qquad \text{(Equation 5-1)}$$

$$V_{RMS} = \frac{V_{peak}}{\sqrt{2}} = \frac{V_{peak}}{1.414} = V_{peak} \times 0.707 \qquad \text{(Equation 5-2)}$$

$$R_{TOTAL} = R_1 + R_2 + R_3 + \ldots + R_n \qquad \text{(Equation 5-3)}$$

$$R_{TOTAL} = \frac{1}{\dfrac{1}{R_1} + \dfrac{1}{R_2} + \dfrac{1}{R_3} + \ldots + \dfrac{1}{R_n}} \qquad \text{(Equation 5-4)}$$

$$R_{TOTAL} = \frac{R_1 \times R_2}{R_1 + R_2} \qquad \text{(Equation 5-5)}$$

$$E = I \times R \qquad \text{(Equation 5-6)}$$

$$I = \frac{E}{R} \qquad \text{(Equation 5-7)}$$

$$R = \frac{E}{I} \qquad \text{(Equation 5-8)}$$

$$P = I \times E \qquad \text{(Equation 5-9)}$$

$$I = \frac{P}{E} \qquad \text{(Equation 5-10)}$$

$$E = \frac{P}{I} \qquad \text{(Equation 5-11)}$$

$$dB = 10 \times \log_{10}\left(\frac{P_2}{P_1}\right) \qquad \text{(Equation 5-12)}$$

$$C_{TOTAL} = C_1 + C_2 + C_3 + \ldots + C_n \qquad \text{(Equation 5-13)}$$

$$C_{TOTAL} = \cfrac{1}{\cfrac{1}{C_1} + \cfrac{1}{C_2} + \cfrac{1}{C_3} + \ldots + \cfrac{1}{C_n}}$$ (Equation 5-14)

$$C_{TOTAL} = \frac{C_1 \times C_2}{C_1 + C_2}$$ (Equation 5-15)

$$L_{TOTAL} = L_1 + L_2 + L_3 + \ldots + L_n$$ (Equation 5-16)

$$L_{TOTAL} = \cfrac{1}{\cfrac{1}{L_1} + \cfrac{1}{L_2} + \cfrac{1}{L_3} + \ldots + \cfrac{1}{L_n}}$$ (Equation 5-17)

$$L_{TOTAL} = \frac{L_1 \times L_2}{L_1 + L_2}$$ (Equation 5-18)

$$E_s = \frac{n_s}{n_p} \times E_p$$ (Equation 5-19)

$$X_C = \frac{1}{2\pi f C}$$ (Equation 5-20)

$$X_L = 2\pi f L$$ (Equation 5-21)

$$Z_p = Z_s \left(\frac{n_p}{n_s} \right)^2$$ (Equation 5-22)

$$\frac{n_p}{n_s} = \sqrt{\frac{Z_p}{Z_s}}$$ (Equation 5-23)

$$\text{Voltage Regulation} = \frac{V_n - V_1}{V_1} \times 100\%$$ (Equation 7-1)

$$P = I \times E$$ (Equation 8-1)

$$I = \frac{E}{R}$$ (Equation 8-2)

$$P = \frac{E}{R} \times E = \frac{E^2}{R}$$ (Equation 8-3)

$$\text{Multiplication factor} = \frac{\text{Transmitter output frequency}}{\text{Oscillator frequency}} \qquad \text{(Equation 8-4)}$$

Oscillator deviation × Multiplication factor = Transmitter deviation

(Equation 8-5)

$$Bw = 2 \times (D + M) \qquad \text{(Equation 8-6)}$$

$$\lambda = \frac{c}{f} \qquad \text{(Equation 9-1)}$$

$$\lambda = \frac{3 \times 10^2}{f\ (\text{megahertz})} = \frac{300}{f\ (\text{MHz})} \qquad \text{(Equation 9-2)}$$

$$\lambda/2 = \frac{150}{f\ (\text{megahertz})} \qquad \text{(Equation 9-3)}$$

$$L(\text{ft}) = \frac{468}{f\ (\text{MHz})} \qquad \text{(Equation 9-4)}$$

$$L_{\text{director}} = L_{\text{driven}}\ (1 - 0.05) \qquad \text{(Equation 9-5)}$$

$$L_{\text{reflector}} = L_{\text{driven}}\ (1 + 0.05) \qquad \text{(Equation 9-6)}$$

$$L(\text{ft}) = \frac{1005}{f\ (\text{MHz})} \qquad \text{(Equation 9-7)}$$

$$L(\text{ft}) = \frac{975}{f\ (\text{MHz})} \qquad \text{(Equation 9-8)}$$

$$L(\text{ft}) = \frac{1030}{f\ (\text{MHz})} \qquad \text{(Equation 9-9)}$$

$$L\ (\text{ft}) = \frac{234}{f\ (\text{MHz})} \qquad \text{(Equation 9-10)}$$

$$L\ (\text{ft}) = \frac{240}{f\ (\text{MHz})} \qquad \text{(Equation 9-11)}$$

$$\lambda\ (\text{ft}) = \frac{984 \times V}{f\ (\text{MHz})} \qquad \text{(Equation 9-12)}$$

$$\text{length (ft)} = \frac{246 \times V}{f \text{ (MHz)}} \qquad \text{(Equation 9-13)}$$

$$\text{SWR} = \frac{E_{max}}{E_{min}} \qquad \text{(Equation 9-14)}$$

$$\text{SWR} = \frac{Z_0}{R} \text{ or SWR} = \frac{R}{Z_0} \qquad \text{(Equation 9-15)}$$

True forward power = Forward power reading − Reflected power reading
(Equation 9-16)

Index